Sexual Selections

MARLENE ZUK

Sexual Selections

What We Can and Can't Learn
about Sex from Animals

UNIVERSITY OF CALIFORNIA PRESS

BERKELEY   LOS ANGELES   LONDON

University of California Press
Berkeley and Los Angeles, California

University of California Press, Ltd.
London, England

First paperback printing 2003

Library of Congress Cataloging-in-Publication Data

Zuk, M. (Marlene)
    Sexual selections : what we can and can't learn about
sex from animals / Marlene Zuk.
        p.   cm.
    Includes bibliographical references (p.   ) and index.
    ISBN 0–520–24075–8 (pbk : alk. paper)
    1. Sexual behavior in animals.   I. Title.

QL761 .Z85 2002
591.56′2—dc21   2001005771

Manufactured in the United States of America

11 10 09 08 07 06 05 04 03
10 9 8 7 6 5 4 3 2 1

The paper used in this publication is both acid-free
and totally chlorine-free (TCF). It meets the minimum
requirements of ANSI/NISO Z39.48-1992 (R 1997)
(*Permanence of Paper*). ∞

CONTENTS

Part Two

UNNATURAL MYTHS

Part Three

HUMAN EVOLUTIONARY PERSPECTIVES

ACKNOWLEDGMENTS

LIKE MOST BOOKS, THIS one was begun well before I knew I would write it, and many people helped it get written. For advice, support, and in some cases merely for responding with interest rather than incredulity to the idea that I was writing a book, I thank Elizabeth Carpelan, David Edwards, Patty Gowaty, Sarah Blaffer Hrdy, Kristine Johnson, Marcy Lawton, Nancy Moran, Virginia Morell, and especially John Rotenberry. Sarah Hrdy read the entire manuscript and made many useful comments and suggestions. Kirk Visscher consulted on the weighty question of whether commercial figs and fig products contain wasps. I thank Barry Farrell for encouraging me to write in Santa Barbara many years ago. Adrian Wenner introduced me to the fallacy of the *scala naturae* and to many other problems in the philosophy of science. Doris Kretschmer of the University of California Press was a thoughtful and insightful reader and editor. Several chapters were written while I was a Visiting Professor in the Department of Animal Ecology at Uppsala University in Sweden, and I am very grateful to the department members for their kindness during my stay and for the support of the Swedish Natural Science Research Council. Other scientists in Finland, Norway, and Sweden graciously discussed their work with me during my visit and gave me access to unpublished material.

My graduate adviser and friend, the evolutionary biologist William D. Hamilton, died before I had a chance to show him this book, which I deeply regret. He taught me a great deal, and was always appreciative of my writing. I wish we could have had the opportunity to talk about the contents. Bill greatly admired A. E. Housman, and the poem that inspired the title for the Introduction was read by his sister Janet Hamilton at his memorial service in Oxford.

NOTE ON SPECIES NAMES

NAMES ARE VERY IMPORTANT to scientists, as they are to many other people, and the exact identification of a particular type of plant or animal can generate a great deal of discussion and occasionally even animosity. One problem with using local names for organisms is that the same creature will have different names in different parts of the world, so what is called a cardinal in Michigan may be called a redbird in parts of the South. Alternatively, the same name, such as "wildcat," may be used for several different kinds of cat. Scientists have dealt with these difficulties by giving each organism two names, in Latin, following a system originated by the Swedish botanist Carl Linnaeus in the eighteenth century. The first name, the genus, may be shared by several similar types. Thus, the white-crowned sparrow is *Zonotrichia leucophrys* and the closely related golden-crowned sparrow is *Zonotrichia atricapilla*. The second part of the name is the species designation, and when combined with the genus name, it serves to uniquely identify the organism as distinct from all other organisms on earth. The genus name is always capitalized, the species name never is, and both are italicized or underlined in print.

I have given the scientific name for every animal mentioned in this book but have not designated higher order nomenclature, such as family or class names. Common names are standardized for some animals, such as birds, and I use these when applicable, although I do not follow the American Ornithologists' Union rules about capitalizing the first letters.

Introduction

AN ODE TO WITLESSNESS

" . . . nature, heartless, witless nature . . ."

A. E. HOUSMAN

SHORTLY AFTER I ENTERED graduate school at the University of Michigan, a fellow student came into my office and flung himself into the chair opposite mine. "I don't understand," he said, "how you can have feminist politics and still be interested in all that stuff over in the museum." The museum was the Museum of Zoology, and the "stuff" to which he referred was the burgeoning field of sociobiology, the study of the evolution of social behavior. It had become a flashpoint for vitriolic debate about the ability of science to draw conclusions about animal behavior in general and human behavior in particular. Both sex, meaning the genetic distinction between male and female, and gender, referring to its social and political associations, were a big part of the controversy from the start. Feminists were quick to recognize that a classic application of biology to oppression had been via the old "anatomy is destiny" route, and sociobiology seemed to some like the same restrictions dressed in trendy new genes.

The debate has taken many turns in the years since; some stereotypes have fallen, and some new perspectives have been achieved. One result of the feminist movement is that many more of the scientific participants are now women. The term "sociobiology" became sufficiently politically laden that it has been abandoned by many scientists, who now tend to call studies of the evolutionary basis of behavior in animals "behavioral ecology" and its counterpart in humans "evolutionary psychology." Yet we are as far as ever from consensus on what feminism and biology have to offer each

I

other and whether—and if so, what—we can legitimately expect to learn about ourselves, particularly about aspects of our sexuality, from studies of nonhuman animal behavior.

I am both a feminist and an evolutionary biologist interested in animal behavior. In my work I am interested in mating behavior and the evolution of sexual characteristics, and I am continually struck with the ways in which our biases about gender influence how we view animal behavior. As a feminist, I advocate the social and political equality of men and women. As an animal behaviorist, I want to learn as much as I can about what the animals I observe are actually doing, and why. In both of these aspects of my identity, I find it impossible to ignore that all of us, scientists, social scientists, and the general public, cannot seem to help relating animal behavior to human behavior. The lens of our own self-interest not only frequently distorts what we see when we look at other animals, it also in important ways determines what we do not see, what we are blind to.

This book is about seeing what animals do. It is about the connections, legitimate and illegitimate, between learning about them and learning about ourselves. It is for those wanting to see how our ideas about sex have helped and hindered our ability to see animals clearly, for those wanting to know about some of the new frontiers in behavioral research, and for those who wonder how we could ever do science without trying to understand our social predisposition. It is for biologists, including those who never thought feminism mattered, and for feminists who always knew it did. I hope to convince you that the natural world is much more interesting and varied than we are often willing to recognize, but that if we try to use animal behavior in a simplistic manner to reflect on human behavior, we will, in myriad ways, misperceive both.

One way we do this is to interpret animal behavior in terms of stereotypical ideas about human society. For example, many feminists have complained about sociobiology's supposed portrayal of females as coy, waiting around for the males to fight it out so they could cheerfully go off with the victor, or at the very least playing hard to get until the sex-mad males had demonstrated which one deserved to win. This image, they claimed, came from outdated and sexist ideas about the nature of women. It is equally true that it is a recipe for being less likely to recognize female assertiveness when it occurs among, say, spiders. The discovery that extra-pair copulations are common in many bird species long thought to be strongly pair-bonded shocked some scientific observers as well as the public; it seemed somehow not just to reflect on, but even to affect our own dubious potential for being monogamous. We both judge these animals by rules for human behavior and at the same time look to them as role models.

We also relate selectively to animals, feeling closer to the cute, fuzzy ones and elevating some species—dolphins and other cetaceans and, more recently, bonobos, formerly known as pygmy chimpanzees—to the status of icons. Why do we love some species more than others? Why is any one species worthy of our concern? E. O. Wilson, the founder of sociobiology, calls the human love of nature "biophilia," a term that has caught on to express our emotional attachment to animals, landscapes, and wilderness. He and others argue, I believe correctly, that tapping into these feelings is essential to efforts to preserve biodiversity. But not only do some animals capture our hearts while others do not; our gender stereotypes confuse this connection, and we create a hierarchy of what should be loved and preserved in nature that can deflect our attention from "lower" species worthy of study in their own right, and can also backfire on former icons in which we lose interest.

We can appreciate dolphins without making them into animal Einsteins, and we can use them in our ongoing struggle to understand intelligence without making them rank above or below other animals. The evolutionary tree is not a hierarchy. It is tempting for all of us to view animals with which we share a more recent common ancestor as being just like us. Baboons and even bluebirds can look and act an awful lot like people. A good deal of my own research is done with insects, and one of the reasons I like working with them rather than with vertebrates is that it is harder to see myself reflected in their behavior. Identification and anthropomorphism are more difficult with insects, and that is a good thing. I do not want to study animals only to learn about me, though that may happen along the way. I want to learn about the insects.

What, then, is the relationship between feminism and the study of gender in other animals? What do feminism and biology have to offer each other? I think the answer is complex. On the one hand, many assumptions about male dominance in nature are falling before contemporary research; being aware of science's past tendency to view males as the only interesting organisms allows us to curtail it. But on the other hand, trying to use science to further a feminist agenda does not serve us or other animals well. Seeking examples of liberated animal females is another example of twisting the natural world into an order it does not show. It blinds us to the variety in animal behavior and involves us in a male-versus-female argument that leads nowhere.

What I advocate is not detachment, nor domination, nor the existence of a special relationship of women with nature. Feminism, however, has more to offer biology than biology has to offer feminism. Feminism provides us with tools to use in the examination of ourselves and other species that can, if we apply them carefully, help us to remove ourselves from the center of things and struggle to see past our biases to what animals are doing.

THE NATURE AND NURTURING OF SOCIOBIOLOGY

The sociobiology controversy, recently expertly analyzed by Ullica Segerstråle in her book *Defenders of the Truth,* is in important ways still with us, despite changes in terminology. The original debate began in the mid-1970s, with the publication of Wilson's *Sociobiology: The New Synthesis.* Wilson, an entomologist by training and avocation, specializing in the study of ants, devoted the vast majority of the book to nonhuman animals. The last chapter, however, speculated about the evolution of human sociality and suggested that aspects of human life such as warfare and a sexual division of labor had biological roots. It was this thin layer of concluding material that sparked all the furor among those worried about the misuse of science in the name of social policy. Exactly what Wilson meant by biological roots is open to interpretation, but his detractors thought he opened the door to a host of politically repressive ideas by supporting existing inequities between the races, classes, and sexes.

Proponents on either side have included some of the heaviest hitters in science, among them the paleontologist and evolutionary biologist Stephen Jay Gould of Harvard (just a floor away from Wilson himself) and Richard Dawkins from Oxford. The battle, which originally pitted mainly left-wing intellectuals and social scientists against more genetically oriented traditional scientists, has had connections to many other debates about the political motivations of scientists and the social implications of what they do. The conflict ranged both wide and deep, harking back in time to the accusation that IQ tests were inherently racist as well as reaching into the "Science Wars" between traditional scientists and scholars from the humanities. The potential for a genetic basis for violent crime and the implications for affirmative action programs have also been part of the argument, with critics maintaining that if we are led to believe that genetics dictate behavior, then social programs designed to prevent children from developing criminal behavior, or to compensate for previous discrimination, are destined to fail.

Both sex and gender were a big part of the sociobiology controversy from the start, for several reasons. If, for example, the pattern of women staying home while men went out and hunted/climbed the corporate ladder was linked to our biology, the criticism went, the women's movement was doomed. Just as nineteenth-century physicians and scientists had claimed to find biological evidence for the intellectual inferiority of women, in either purported differences in brain size, the demands of menstruation and childbearing, or muscular frailty, so their modern counterparts seemed to be suggesting that evolutionary tendencies shaped hundreds of thousands of years ago made women coy, uninterested in sex, and unwilling to take risks, whether on the playing field or in the stock market. Numerous feminist theorists, including some scientists, such as Anne Fausto-Sterling, a developmental biologist at Brown University, attacked sociobiology as sexist claptrap thinly veiled as science.

Sex also figures in the debate for the simple reason that sex—simple sex, as well as gender—is an integral part of evolution. Anyone explaining the evolution of behavior, particularly in animals but to an arguable extent in people as well, is mainly concerned with two things: food and sex. Natural selection occurs through the differential reproduction of individuals; variants with better abilities to keep warm, resist disease, and fend off predators will leave more offspring, who in turn can also do these things better, than other variants. Food is important because without it organisms cannot live long enough to reproduce, and sex is important because without it most organisms, by definition, do not reproduce at all. One could argue, in fact, that food is important only in the context of sex, since an animal that successfully locates all the ripe fruit in the forest but fails to mate is an evolutionary dead end.

THE POWER TO CHARM

This part of sex is, however, only the most obvious reason for its significance in evolutionary biology. The more subtle explanation is called sexual selection, and it was developed as a theory to account for differences between males and females, both morphological and behavioral, that seem removed from the immediate necessities of reproduction. Like the idea of natural selection, sexual selection theory is widely accepted among biologists, and also like natural selection, sexual selection has its origin in the work of Charles Darwin.

When Darwin began to develop his ideas about the origin of species, he distinguished between traits used for survival and those used in acquiring mates. He pointed out that while many animals exhibit extreme traits, in some cases these are found in both sexes and turn out to be beneficial in daily life, like the elongated curved bills of Hawaiian honeycreepers, which are used for probing flowers for nectar. Other extreme traits, though, are sex-limited, and Darwin devoted an entire book, *The Descent of Man and Selection in Relation to Sex,* published in 1871, to explaining them, noting that many of the characteristics seem actually detrimental to survival. In several of the species of birds of paradise, for instance, the male has ornamental feathers so long or elaborate that they impede his flying ability.

Darwin also distinguished between traits such as these, which are strictly speaking not needed to reproduce, and what he called the primary sexual characters—the plumbing, so to speak, that makes males able to produce sperm and females able to produce and nurture eggs. He figured that a trait allowing a female to put a water-resistant shell around an egg, for example, would be unequivocally beneficial to her, and fit under the general category of natural selection. But what about the other traits, the long tails and bright colors and structures like antlers on deer? Darwin called those traits secondary sexual characters, and noted that in many cases they simply could not seem to have arisen through natural selection. A brightly colored set of feathers or a loud song probably makes a male more conspicuous to predators, and either may be physiologically costly to produce. How could the bearers of the traits have been favored by selection over their less elaborated counterparts?

Darwin said that sexual selection, a process similar to but distinct from natural selection, had led to their evolution. The secondary sexual characters could evolve in one of two ways. First, they could be useful to one sex, usually males, in fighting for access to members of the other. Hence, the antlers and horns on male ungulates, like bighorn sheep, or on the aptly named male rhinoceros beetles. These are weapons, and they are advantageous because better fighters get more mates and have more offspring. The second way was more problematic. Darwin noted that females often pay attention to traits like long tails and elaborate plumage during courtship, and he concluded that the traits evolved because the females preferred them. Peahens find males with long tails attractive, just as we do. In one of my favorite passages from *The Descent of Man* Darwin marvels, "We shall further see, and this could never have been anticipated, that the power to charm the female has been in some few instances more important than the power to conquer other males in battle" (p. 583). The sexual selection process, then, consisted of two components: male-male competition, which results in weapons, and female choice, which results in ornaments.

While competition among males for the rights to mate with a female seemed reasonable enough to Darwin's Victorian contemporaries, virtually none of them could swallow the idea that females—of any species, but especially the so-called dumb animals—could possibly do anything so complex as discriminating between males with slightly different plumage colors. Alfred Russel Wallace, who independently arrived at some of the same conclusions about evolution and natural selection that Darwin did, was particularly vehement in his objections. He, and many others, simply found it absurd that females could make the sort of complex aesthetic decision required by Darwin's theory. After all, according to the thinking of his time, even among humans only those of the upper social classes could appreciate aesthetic things like art and music, so it seemed ridiculous to imagine that animals could do something many humans—particularly non-Englishmen—could not. Several authors have also suggested that because females were not supposed to be interested in sex anyway, the idea that they spent time thinking about it made Victorian scientists uncomfortable. Besides, what would be the point of choosing one male over another? If the only difference between them was the secondary sexual trait, why should the female bother? Wallace scoffed, "A young man, when courting, brushes or curls his hair, and has his moustache, beard or whiskers in perfect order, and no doubt his sweetheart admires them; but this does not prove that she marries him on account of ornaments, still less that hair, beard, whiskers and moustache were developed by the continued preference of the female sex" (p. 286).

Largely because of this opposition to the idea of female choice, sexual selection as a theory lay dormant for several decades. The work of the British geneticist R. A. Fisher was a notable exception, but in general even after genetics became incorporated with Darwin's ideas on evolution to form what is called the New Synthesis, the major evolutionary biologists of the early twentieth century—George Gaylord Simpson, Theodosius Dobzhansky, Robert Ledyard Stebbins, and their contemporaries—were largely uninterested in sexual selection. When they discussed extravagant traits at all, they suggested that these arose to allow females to find a mate of the right species. Choosing a male of a different species could have disastrous consequences, because hybrid offspring, if they can develop at all, are often infertile. In general, variation among individuals was not seen as particularly interesting, so long as reproduction continued.

It was not until the 1960s that evolutionary biologists began to reconsider the portrait they had painted of animal social life. Suddenly, it seemed, people realized that males spent an awful lot of time showing off to females during the breeding season, and it became increasingly hard to believe that all the fuss was made merely so that a female cardinal could tell the difference between a male cardinal and a duck.

It would be interesting to speculate about the social and cultural forces that led scientists to reevaluate their views on sexual behavior. Within the field, however, probably the most important new insight came from a paper written by the evolutionary biologist Robert Trivers about thirty years ago. He pointed out that in many species, females and males inherently differ because of how they put resources and effort into the next generation. Females are limited by the number of offspring they can successfully produce and rear. Because they are the sex that supplies the nutrient-rich egg, and often the sex that cares for the young, they have an upper limit set at a relatively low number. They leave the most genes in the next generation by having the highest quality young they can. Which male they mate with can be very important, because a mistake in the form of poor genes or no help with the young can mean that they have lost their whole breeding effort for an entire year. Males, on the other hand, can leave the most genes in the next generation by fertilizing as many females as possible. Because each mating requires relatively little investment from him, a male that mates with many females sires many more young than a male mating with only one female. Hence, males are expected to compete among themselves for access to females, and females are expected to be choosy, and to mate with the best possible male they can.

This, of course, should sound familiar: it is the same division of sexual selection that Darwin originally proposed. But Trivers not only gave it a new rationale. What he did in addition was to bring female choice back to the forefront of sexual selection, and suggest a more modern underlying advantage to it—even though he and others often referred to females as "coy," with the implication that the impetus for sex came largely from males, who fought among themselves to get to the females and allow the choices to occur. Furthermore, ideas about the evolution of behavior had advanced enough that scientists no longer worried about an "aesthetic sense" in animals; it didn't matter how females recognized particular males, just that if they did, and it was beneficial, the genes associated with the trait females were attracted to would become more common in the population than the genes of less-preferred traits. Evolutionary biologists, therefore, could ignore questions about motivation and get to the more testable issue of how discrimination among males might result in the evolution of ornamental traits that did not function either in day-to-day life or in male combat. Female choice made sense.

Current work on female behavior in many species of animals has confirmed Trivers's—and Darwin's—basic idea about female preference for particular types of males being a major force in evolution. Again and again, females have been shown to be able to distinguish small differences among available mates, and to prefer to mate with those individuals bearing the most exaggerated characters. In some cases those males are also more healthy and vigorous, so that ornaments appear to indicate not just attractiveness but the ability to survive. Peacocks, often used as the symbol of sexual selection, provide one of the best-known examples. The British biologist Marion Petrie studied the behavior of flocks of peafowl that were allowed to range freely in a park in England. She discovered that females did indeed prefer males with greater numbers of eyespots on their tail feathers, and that this preference could be manipulated by cutting the eyespots off of some males' tails; females lost interest in the pruned peacocks and became attracted to the untrimmed ones. Even more interesting, she allowed females to mate with males that had variable numbers of eyespots, and then reared all the offspring in communal incubators to control for differences in maternal care. The chicks fathered by the more ornamented males weighed more than the other chicks, an attribute usually connected with better survival in birds. Indeed, when the individually marked chicks were then released into the park and recaptured the following year, the ones with the more attractive fathers also were found to be more likely to evade predators and survive in the semi-natural conditions.

Not all cases are so satisfyingly clear-cut, but modern biologists accept female choice as an important part of sexual selection. What about the accompanying notion that females were therefore coy, uninterested in sexual activity unless it was initiated by the ever-eager males? This has not fared so well. Evidence from insects, birds, primates and other organisms has contradicted the idea of the passive female and suggests instead that females often mate many times, with many different males. Nevertheless, the basic principle that males are limited by the number of eggs they can fertilize (which can potentially be very high) while females are limited by the number of offspring they can produce and, if necessary, rear (which is potentially relatively low), is a general one that leads to differences between the sexes. Sometimes, if males invest a great deal in offspring along with females, these differences will be quite small; sometimes they will be quite large. How the differences are interpreted is another story, and one that forms the basis for this book.

GENES: SELFISH, SEXY, OR MISUNDERSTOOD?

Sexual selection research has become one of the hottest areas in evolutionary and behavioral biology. Scientists have found enormous variation in Darwin's original scheme, with both males and females behaving in ways that go far beyond Victorian stereotypes. The field has never been without its critics, however, and the criticisms have been made on both social and scientific grounds, with the distinction between the two often blurring. These criticisms were in part what led to my fellow graduate student's assumption that my feminism and my science must necessarily be at odds.

I have never found any basic conflict between my belief in sexual egalitarianism and my interest in sexual behavior among animals, including my endorsement of the theory of sexual selection. Whatever Darwin's personal views on women, he had managed to hit on an enduring concept in biology that has not appeared to depend on one's political views to hold up.

How, then, do feminism and attempts to use evolutionary theory to explain behavior interact? As I mentioned above, one immediate reaction from some was that so-called biological explanations have so often been used to justify unequal treatment of groups, including males and females, that any new efforts should be viewed with suspicion. The critics focused particularly on efforts to apply evolutionary theory to human behavior, but all links between behavior and selection were often seen as tarred with the same brush. Here I will briefly discuss some of the common misconceptions about evolution and behavior as they apply to the controversy.

First, many people are leery of the apparent consciousness attributed to animals and, at times, their genes, during the process of evolution. The idea of female "choice" still suggests a conscious weighing of alternatives, an idea that seems anthropomorphic at best and idiotic at worst when applied to animals, particularly invertebrates, such as insects, which lack sophisticated brain components traditionally associated with decision-making in humans. Even for humans, the idea has been called into question for social reasons; Segerstråle notes (p. 172) that the anthropologist Edmund Leach decried "this curious idea that by and large individuals can somehow choose their mates! In most of the world they can't! Their love affairs are different from their marriages. Their marriages are arranged by their seniors for political reasons."

For evolutionary biologists, however, the process is not as important as the consequences. Selection acts only indirectly on mechanisms, if it can be said to act upon them at all. If we can show a relationship between a trait and a female tendency to mate with those bearing it, sexual selection may be operating. If female beetles, when presented with one male bearing two spots on his back and one with four spots, are more likely to mate with the four-spotted variety, more baby beetles that develop four spots as adults will result. Two-spotted beetles will become less frequent in the population, and, on the assumption that spottiness has no relation to survival, sexual selection via female choice will have caused the evolution of a secondary sexual character, spot number. Although it would be interesting to know the mechanism by which females discriminate among prospective mates, and this has relevance for formulating some models of preference, it does not matter for the sheer demonstration of female choice what went on in the nervous system of the female, much less that she is incapable of formulating a rational thought. Even with humans, what goes on in the mind is often less significant than what results from the behavior. This is not to suggest that studying sexual selection in either humans or animals, but particularly the former, is without problems. We need not, however, confuse conscious decisions with evolutionary outcomes.

The next misconception concerns the related specter that then rears its head: the nature of genetic differences in behavior, a necessary precursor for selection to act on those differences. What does it mean for a behavior to "be genetic"? Does it mean that possession of a particular form of a gene always leads to the execution of a particular behavior? Does it mean merely that the potential for the behavior is there? Here the relationship between mechanism (getting from genes that produce proteins to a response in the nervous system to a stimulus) and consequence (perhaps changes in fertility or attraction to mates of a certain type) is even more difficult. We have known for many years that genetic differences alter behavior, even fairly complex behavior, and most medical practitioners now recognize, for example, that many mental illnesses have a genetic component. Yet the field of behavior genetics, even as applied to nonhumans, has had an uneasy history, haunted by the eugenics movement, unable to shake the accusation of genetic determinism, of suggesting that if genes influence behavior, they must perforce dictate behavior. This is a misconception about the way genes interact with their environment to produce a trait. One misunderstanding has led to another, as the notion of genes dictating behavior segues into what is called the "naturalistic fallacy," the idea that what is natural is good, so if behavior is genetic, and genes are part of our nature, then we can all give up on trying to change the world into a more just place. Finally, arguments have raged about whether such traits as homosexuality or altruism are "genetic or learned," "innate or culturally determined," due to "nature or nurture."

I discuss the inherent problems with the nature-nurture dichotomy in Chapter 3, in the context of the maternal instinct. Suffice it to say here that all behaviors are the result of genes, developmental conditions during embryonic life, and the subsequent environment in which the organism finds itself. If two genetically identical organisms experience different environments, and exhibit two different manifestations of a behavior, one can conclude that the difference is due to the environment. Conversely, if two genetically dissimilar individuals experience the exact same environment, and still show differences in behavior, one can conclude that genes cause the difference. What can be said to be genetic or learned is a difference in a trait, and not the trait as such. Difficulties with actually putting this distinction to a test notwithstanding, it points up the absurdity of arguing over which part of a behavior, whether it is hole-drilling in woodpeckers or homosexuality in humans, is innate or cultural. This is not to say that we can airily dismiss concerns over the influence of the environment and assert that genes are the only subject of interest, any more than we can say that all human behavior is cultural and hence evolution is of little relevance.

Nowhere is this unease about genetic explanations of behavior more apparent than in attempts to explicitly account for the evolution of how we humans behave. Some critics, not just of sociobiology but of scientific approaches to human biology in general, have objected to the idea that people, with our flexible behavior patterns and extensive period of childhood learning, could be considered as just another species. One found such an assumption "arrogant," which is a curious reversal of the more frequent suggestion that it is special pleading to argue that humans have a separate exalted place in nature. Others simply find social and sexual behavior—sometimes all such behavior, sometimes only when it occurs in humans—to be so complex that we cannot ever guess its trajectory through evolutionary time.

My own concern with this problem of humans being "special" takes us back to the sociobiology controversy and feminism. I am perfectly ready to accept that humans are subject to selection in the same way as other organisms, which places me squarely in the sociobiology camp. On the other hand, I recognize that self-awareness, which is so highly evolved in humans, necessarily complicates matters. If one agrees that evolution affects our behavior, then one must surely also agree that evolution influences how we view ourselves, a catch-22 if ever there was one. Self-consciousness allows us to examine our behaviors (as well as those of other animals), but the way we interpret those behaviors influences our abilities to see them clearly.

It is not news that humans selectively look at the world, both their own and that of other organisms. One of the great contributions of the science of animal behavior has been to point out the dangers of such selectivity, particularly when combined with anthropomorphism. A favorite example of mine which illustrates the problem comes from E. L. Thorndike, an animal psychologist at the turn of the twentieth century who formalized the systematic, experimental study of behavior. In a monograph published in 1898, he rather peevishly took to task previous attempts to examine the mental processes of nonhumans. He wrote:

> In the first place, most of the books do not give us a psychology, but rather a *eulogy* of animals. They have all been about animal *intelligence,* never about animal *stupidity.* . . . In the second place the facts have generally been derived from anecdotes. . . . Besides commonly misstating what facts they report, they report only such facts as show the animal at his best. Dogs get lost hundreds of times and no one ever notices it or sends an account of it to a scientific magazine. But let one find his way from Brooklyn to Yonkers and the fact immediately becomes a circulating anecdote. Thousands of cats on thousands of occasions sit helplessly yowling, and no one takes thought of it or writes to his friend, the professor; but let one cat claw at the knob of a door supposedly as a signal to be let out, and straightway this cat becomes the representative of the cat-mind in all the books (p. 4).

This problem has of course persisted in science, and I will explore its ramifications as they pertain to sexual behavior in several of the following chapters. In the meantime, Thorndike's complaint can quite easily be reworded to reflect ideas about sex roles; if, for example, someone finds that female rabbits or tortoises or houseflies are less active than males, this reinforces stereotypes about passive females, whereas if they discover the reverse, less notice is taken. Furthermore, people may be less likely to notice behavior in the first place if it contradicts a stereotype. As the psychologist Virginia Valian has pointed out, we interpret what we see in terms of "gender schema," ideas about what the sexes are like, physically, mentally, and emotionally. If men are generally viewed as tall, we see them as tall, and tests show that people overestimate height of men and underestimate that of women. If men are generally viewed as capable and authoritative, we will see them that way, too, whereas if women are stereotyped as submissive and incompetent, we will tend to judge them that way even given evidence to the contrary. The result has obvious implications for practical issues like the salaries of men and women in the same occupation, but it also colors our ability to interpret or even detect the behavior of other species as well as humans.

Does rejecting such stereotypes mean rejecting evolutionary explanations of behavior? I do not believe it does. The question is not whether we accept biological explanations or reject them, it is how much and in what ways the explanations suffer from our biases.

WITLESS NATURE

According to Segerstråle, both E. O. Wilson and Konrad Lorenz, the Nobel Prize–winning ethologist who developed the notion of young imprinting on their parents, were proponents of the naturalistic fallacy, that what is natural is good. Both felt that universal laws about morality in human behavior arose from the working of nature. Both were concerned that inattention to our evolutionary history could contribute to nuclear war or other catastrophes. This attitude does not, however, automatically arise from an evolutionary perspective on behavior. It is also true that examining nature with an eye toward our human tendency to force it to say certain things can be enlightening all by itself.

A way out of the dilemma concerning the relationship of stereotypes and evolutionary explanations of behavior simultaneously provides a solution to the naturalistic fallacy. It is perhaps best stated in a poem by

A. E. Housman, an early twentieth-century Englishman described as a "Romantic pessimist" who is often read in high school literature classes but does not usually serve as a source for information about philosophy of science. The poem, from his *Last Poems,* is in many ways a celebration of knowing nature, of seeing:

> Where over elmy plains the highway
> Would mount the hills and shine,
> And full of shade the pillared forest
> Would murmur and be mine.

It ends with a verse that summarizes a remarkably evolutionary view of the world:

> For nature, heartless, witless nature,
> Will neither care nor know
> What stranger's feet may find the meadow
> And trespass there and go,
> Nor ask amid the dews of morning
> If they are mine or no.

Nature, as he says, is witless. It is not kind, not cruel, not red in tooth and claw, nor benign in its ministrations. It is utterly, absolutely impartial. I myself take this in the most positive possible way, finding it restful that the world comes without an agenda. This does not mean we cannot have our own agendas, just that we cannot claim that ours has been lifted from some higher outside source. Further, witlessness is not at all the same thing as stupidity. It simply suggests that we cannot expect to find a user's manual accompanying the actions of animals. What is natural can't be inherently "good" any more than it can be inherently amusing, or inherently painful. Finding out that some animals kill their young says no more about the ethics of infanticide than finding out that some animals are yellow says about fashion trends.

Witlessness can, however, be extraordinarily illuminating. When we begin to understand the details of animals' lives, the ways in which we have been trying to make generalizations about behavior, about sex roles as well as selfishness, suddenly seem peculiar and useless. It is as if we were embarking for a space station with elaborate plans for improving the design of a sailing vessel or, perhaps, as if we were blasting off with plans for improving soufflés. Nature does not provide object lessons so much as challenges to our assumptions. This is not to say that we can never generalize, because science relies on generality, but that the generalizations need to come from a wider base. To answer my graduate student friend, I can do what I do because nature is witless, in the sense of being impartial. Feminist points of view can help us look at science from a different angle, but they will never be able to change nature, something for which we can all be grateful.

In this book I try to show that although looking at nature can result in different interpretations, this does not mean that all attempts to study the world are just culturally derived exercises relevant only in a certain social context, the way some philosophers and social scientists might have us believe. It is nonetheless true that we and our culture and our history throw up different kinds of barriers to seeing clearly, especially where sex and gender are concerned. How can feminism help? It can give us some tools to use in the examination.

The chapters in Part I examine various sorts of biases with which we often color the world we are looking at and ask in what ways a feminist perspective might make things appear otherwise. Here I am concerned with how stereotypes distort the questions we ask as well as how we answer them. Feminists have identified several ways in which scientists, by taking males as the norm, have limited our views of what females do, and I explore these. Male bias, however, is far from the whole story, and some attempts to counter it lead in unfruitful directions. I therefore also examine useful and nonuseful modes of attacking stereotypes.

Part II is concerned with myths that, on a deeper level than biases, prevent us from seeing what animals do. The principal issue here is that of the *scala naturae,* the hierarchical view of the natural world, and particularly the animal world, that has long been deeply embedded in Western thought and still informs many aspects of our ways of thinking. The history of how humans have viewed our place in nature comes with baggage that we barely realize we have. Spinning off from it are myths that blind us in more particular ways—about kinship, about communication, about dominance—and here too I ask how turning a feminist light on the inquiry can improve our vision.

Part III specifically takes up four aspects of human behavior—female orgasm, menstruation, homosexuality, and spatial ability—and explores their relationship to evolution, asking whether or to what extent they represent adaptations, as opposed to by-products of selection for some other process, or what we can say about how and why they developed. I examine a range of views as to their possible adaptive significance and the state of research into parallel behaviors—and the lack of them—in non-human animals, attempting to assess what we can learn about ourselves from these findings and what is likely to lead us into blind alleys.

The final chapter describes some of the ways animal behavior can be misused in discussions of gender by both "sides" of the battle of the sexes. Though I believe that feminism has more to say to biology than biology does to feminism, I conclude by discussing the role of biology in understanding sex differences and similarities, and suggest that biology can extend the boundaries of our thinking about gender as it can for so many other ideas. Contrary to popular belief, biology does not set limits, it demolishes them.

I hope that readers will feel stimulated to pursue in greater detail some of the topics I discuss. To that end, for each chapter I have listed selected readings that are gathered at the end of the book; these are intended to steer the interested reader toward some of the original scientific papers as well as more popular books and articles. I annotate each reading with a brief description of its contents. The list is not intended to be exhaustive but should serve as a jumping-off point for each general theme.

PART ONE

Sexual Stereotypes and the Biases That Bind

One

HE STOOD OUTSIDE THE DOOR of the museum, barefoot, very tan, and wearing only a faded pair of denim cut-offs. He was clutching a large bird to his chest. The bird was barely alive, eyes shut, its black and white feathers moving slightly. "Can you help him?" asked the young man, staring hopefully at me. After finishing my undergraduate degree, I worked in the small vertebrate museum on the campus of the University of California at Santa Barbara, and such requests were not uncommon. Birds frequently washed up on the beaches covered in oil or otherwise hurt, and sometimes we could let them heal in a cage in the museum until they were ready to fly away.

Not this time, however. I pointed out that the bird, which I identified as a common loon *(Gavia immer),* was nearly dead, and that while there was nothing to be done for it, I'd be happy to take it for the museum's skin collection. Stuffed birds were used both as research specimens and for teaching; in my vertebrate zoology course I had learned how to identify many of the local species by painstaking examination of such taxidermy mounts. Now, having graduated, I was learning how to prepare the specimens myself, and fresh material was always welcome.

He was horrified. How could I be so ghoulish when the poor creature was still alive? While we were arguing about it, the loon died in his arms. I eyed it cheerfully. After a little more persuasion, he agreed to donate the bird to our collection, and I started the paperwork. Skins are always more valuable if information about their collection is kept with the specimen, so I noted the date, the place on the beach where it had been found, and then asked the man his name.

"Wing Bamboo," he said.

I paused. It was southern California, it was the '70s, but while people named Rainbow and Runningwater were commonplace in the food co-op, you usually didn't see them in the museum donating dead birds. Should I write, "Bamboo, Wing"? "W. Bamboo?" Was it all one word? In the end I wrote it down just as he'd said it, and told him to put the bird on the table. He gazed at it and put it down, but only after clutching it a little tighter and intoning, "Goodbye, Brother Loon."

At the time I just rolled my eyes and put the bird in a plastic bag with a label, figuring he'd seen too many reruns of the then-popular film *Brother Sun, Sister Moon*. But since then I have thought about the encounter many times, and two elements of it remain intriguing. First, why did he need to claim kinship with a bird that he could not identify, knew little about, and had never interacted with? Second, why "brother" and not "sister"?

ANIMAL ROLE MODELS

People have always looked to animals as illustrations and models of behavior. From Aesop to the Bible to modern literature, animals have been held up as representing virtues and vices: industrious bees and ants, wily foxes and untrustworthy snakes. It is only a short leap from these fables to the conviction that certain types of behavior in humans and nonhumans alike are natural, meant to be the way they are. Perhaps because we hold our ideas about sexuality and gender very dear, nowhere is this claiming of biologically intuitive high ground more clear than in matters of sex and sex roles. Watch a mother bird bringing a beak full of insects to her nestlings, or better yet a mother baboon with her infant on her back. It is easy to conclude that females must instinctively know how to be mothers, so maternal behavior is natural and voluntary childlessness is not. Production of offspring in most animals requires expenditure of effort by a male-female pair; by definition, sexual reproduction requires sperm to meet egg. Thus, exclusively same-sex pairing is not widely seen in wild animals, and thus again it is easy, or at least tempting, to conclude that heterosexuality is natural, homosexuality is not.

Another example: in the vast majority of animals, overt aggression is more likely to be observed in males. Impressive weaponry like horns and antlers are generally seen, or at least are best developed, in male ungulates, not females. Although the concept of the pecking order was originated in the early 1900s by the Norwegian behavioral biologist Thorleif Schjelde-rup-Ebbe after watching groups of female chickens, we tend to associate the pecking order and its relative, the dominance hierarchy, with males, and the term "alpha male" is part of our everyday vocabulary. The leader of the pack is not supposed to be a girl, or presumably even a female wolf, whatever evidence exists to the contrary. Another small step takes us to the conviction that women, because of their naturally nurturing ways, should be able to stop wars, help establish the global village, and come up, somehow, with a world vision free from aggression and violence. We use animals as role models, in an odd parody of art imitating life imitating art. We see our stereotypes played out in animals, and then approve or disapprove of human behavior based on whether it fits the roles that we assigned in the first place.

In one sense, there is nothing wrong with this attitude. At least some of the time, we derive comfort from our connections with other animals. People of many cultures use animals as totems, as symbols of characteristics they admire. I like loons, too, and on further consideration I wonder if I am as dissimilar from Wing Bamboo as I originally believed when I scoffed at his farewell and got out a plastic bag. He wanted the loon to be his brother, maybe partly because of some New Age spiritual trendiness, but also partly because it is nice to think that animals, too, live and love and have babies. Cartoons, children's stories, and many nature documentaries are firmly rooted in this belief. Scientists may even encourage such empathy, misplaced or not, because it makes the public more likely to support research into the habits of the animals themselves, and to be receptive to using its results to solve human problems. Viewing animals as caricatures of people is anthropomorphism, of course, and scientists already know the danger it carries of clouding our ability to interpret nature. Beyond that, however, lies a deeper and perhaps more serious problem: the risk that if we claim our kinship too insistently we will not see what the animals actually do, because we will see only behaviors that reflect our own preconceived ideas.

MODEL SYSTEMS: FRIEND OR FOE?

This concern about loss of objectivity is not new, but I want to take it a step further than usual. Scientists use animals as object lessons, too, but instead of calling them, either implicitly or explicitly, role models, they call them model systems. A model system is one that is used to obtain general results about some aspect of biology. For example, the small fly *Drosophila melanogaster* and other species in the genus have been used to study virtually every aspect of genetics possible. (Entomologists, the scientists who study insects, are tediously quick to point out that although commonly referred to as "fruit flies," *Drosophila* are more correctly called "vinegar flies" or "pomace flies." It seems to me that since by definition common names are the ones people regularly use, "fruit fly" is at least as justifiable a name, but entomologists are notoriously hard-headed about such things.) The flies breed quickly, with a generation time of ten days, which facilitates tracing the inheritance of traits; they have chromosomal characteristics that allow relatively easy mapping of genes; and they display a remarkable variety of colors, shapes, and sizes, many of which can be shown to result from a tiny alteration in the chemical structure of the genes. They are also easy to raise in large numbers, subsisting quite happily on a pasty mixture of sugars and yeast in glass or plastic tubes.

But the most important reason that so many scientists study *Drosophila* is that so many other scientists study it as well. The great geneticist Thomas Hunt Morgan suggested its use in 1909, and biology has never been the same. He suggested it for practical reasons, but these were soon subsumed in the snowballing weight of information that made each successive study able to rely on its predecessors. Every life science student has a nodding acquaintance with fruit fly genetics, and I suspect many readers can recall struggles in a high school laboratory to anesthetize flies so they could be viewed under a microscope. Moreover, the work is ongoing. Similar to the Human Genome Project, the European *Drosophila* Genome Project, an effort involving researchers from many labs and almost as many countries, has recently achieved its objective of characterizing every gene on the sex chromosome of *Drosophila melanogaster,* a project that has far-reaching implications for understanding many basic questions in biology. If you want, for your research, a strain of fruit flies with yellow eyes, long legs, fused body parts, sluggish behavior, or exceptionally long sperm cells, you can find it easily through the Internet. Or you can examine FlyBrain, an online atlas and database of the *Drosophila* nervous system. There are even online discussion groups (not quite chat rooms, but close) that are solely concerned with *Drosophila* biology. This wealth of knowledge makes advances in science—not just about *Drosophila,* but about the genetics of many other species, including humans—much easier, because the basic groundwork has been laid and the techniques established.

Drosophila is perhaps the quintessential example of a model system, but the study of animal behavior has its own equivalents. Laboratory rats are classic subjects for many aspects of behavior, including sexual behavior. Researchers have meticulously documented the details of rat mating and reproduction with a level of detail that could bore even the most prurient; for instance, the number of intromissions or pelvic thrusts characteristic of copulating males under a wide variety of environmental circumstances and under the influence of many hormones and drugs is well established. In females, the neural pathways causing lordosis, the back-arching response required for successful copulation, have been studied in more detail than in any other organism, including humans. Again, while rats are convenient in several respects, as in their willingness to live in small plastic cages, eat dry rat pellets, and allow scientists to observe them without becoming perturbed, the biggest reason for using them to study sex is that everyone else does too.

Some of the reasons for using particular species as model systems are historical. If someone back in the early twentieth century worked out how much male rat urine on a piece of cotton is needed to influence the likelihood that a female will come into heat, or estrus, then that is one less thing that a modern scientist needs to establish before continuing with his or her own research. This baseline information may prove useful for studies aimed at a variety of goals, perhaps toward understanding where in the brain olfactory signals are processed so that they can influence later behavior. It does not matter why that person used rats, or whether indeed jerboas, small desert rodents that also survive well in the laboratory, would have been better. If the early scientist had worked on armadillos or musk oxen instead, it wouldn't have mattered, at least from this particular perspective. In fact, a few oddball model species exist, such as water shrews for the study of hormonal influences on behavior and cotton rats for the study of lactation. Certainly practicality enters into it; it is hard to imagine anyone seriously attempting to use musk oxen as the equivalent of lab rats. But in many cases there is no a priori reason to choose one species of fly over another, except to use what everyone else is already using.

Our understanding of behavior is much better for species studied in the laboratory than for those in the wild. Even in the field, however, much more information is available about some species than others, often through the efforts of one or a few key researchers. Sometimes this is because the biologists simply pitched on a species they were able to watch for long periods of time, such as red deer (*Cervus elaphus,* also known as "elk" in North America) studied by Tim Clutton-Brock of Cambridge University, or European barn swallows *(Hirundo rustica),* whose short lives have been documented for many generations by Danish scientist Anders Pape Møller, now at the Université Pierre et Marie Curie in Paris.

In both these examples, the choice was not entirely coincidence; red deer occur in relatively confined populations on Rhum, an island off the coast of Scotland, which makes keeping track of births, deaths, and dominance much easier than it would be in a herd that wandered over great distances and became impossible to follow during dispersal. On Rhum the sex ratio of offspring of individual hinds, or female deer, has been calculated over several seasons, which allowed the testing of hypotheses about the relationship between environmental conditions and the sex of a mother's young. Most people take for granted that most species produce equal numbers of males and females, but the question of how this ratio evolved has been of interest to scientists for many years. The evolutionary biologist Robert Trivers suggested in the 1970s that females should be more likely to produce sons than daughters when the females themselves are in good physiological condition, because males are generally highly competitive and only males of the highest quality are expected to be successful in obtaining a mate. Producing a weakling son is therefore unlikely to yield any reproductive payoff for the mother, whereas a daughter even in poor condition will almost certainly be fertilized by a male and produce grandchildren, the only profit that is significant from the standpoint of evolution. On the other hand, if one's son is very successful in combat with other males, he can potentially sire many more offspring than a single female, even one in the best condition.

While logical, this idea has been very difficult to test. Clutton-Brock and his colleagues have been able to plot the sex ratio of offspring of different red deer females for many years, some when environmental circumstances were luxurious, others when resources were scarce. It now appears that Trivers was right, at least for this species under many conditions and probably for others with similar social systems. Interestingly, the proportion of males born each year declined with increasing population density and heavy winter rainfall, both of which are stressful for pregnant females. This change in the sex ratio was seen at a time when fewer females overall were having calves, suggesting that the difference was caused by differential fetal loss, rather than selective fertilization of eggs. Exactly how the females assess their own condition and then alter, unconsciously of course, the likelihood of conceiving or producing a son or a daughter remains a mystery.

The barn swallows nest readily in wooden boxes constructed and placed in a convenient location by the investigator, and large numbers of them can be captured relatively easily in nets. We know, therefore, that male swallows with longer tail streamers tend to have fewer fleas and other external parasites than males with shorter tails, and that females prefer to mate with such males. Other reasonably well-studied natural systems include North American prairie dogs, song sparrows, and a few of the African carnivores, such as lions and hyenas. For none of these do we even approach the degree of understanding we have of what makes lab rats or *Drosophila* work, particularly in terms of their genetics, but we have some idea about the forces that govern their lives in evolutionary terms, that is, what makes an individual likely to produce more or fewer offspring.

Model systems are a good thing, aren't they? Yes and no. They are a good thing because it would be frustrating in the extreme to have only a rudimentary amount of information about a large number of species, so that we might know a little bit about the mother-infant relationships of seventy-five species of primates but nothing about the adolescence or adulthood of any of them. If we had to reinvent the wheel every time we wanted to study a question about behavior in any species, we would still not have reached the stage where we could, for example, reliably predict the outcome of a dominance interaction between two red deer stags. They are also a good thing because armed with a detailed knowledge of a few species, we can have some hope of coming up with generalizations that will apply, with a few modifications, to many other species—the true definition of a model system. Few people study rats because they are interested in rats per se; they are interested in how to describe general principles of learning, or in the factors governing nerve control of penile erections, or in the inheritance of aggression during early adulthood. This ability to transcend the specific case at hand and make a statement about the way an entire, if small, piece of the universe functions is what science is all about, and it is the strength of the model system.

Many of us, of course, are in fact interested in our study animal for its own sake, and spend long hours watching pandas or magpies because pandas and magpies capture our imaginations, appeal to the same sense of belonging that Wing Bamboo had with his loon. It would be difficult to maintain the enthusiasm necessary for the hard work of doing science if one felt indifferent to the animals themselves, and no one denies the inherent appeal of watching many animals in their natural habitats. And certainly one scientist's charming Bambi is another's noxious pest; I admit to a personal fondness for earwigs that I know is not shared by most people, professionals and amateurs alike. I think they are cute and interesting, whereas almost everyone else, including the students I have tried in vain to convince to work on them, finds earwigs about as endearing as income tax. But interest in natural history, while it may provide the impetus for beginning or sticking with a project, does not form the basis of the scientific study of animal behavior. For that we still rely on model systems.

When are model systems bad things? They are bad things when they cease being convenient test cases for theory and begin to be models in the sense that a layperson uses the term. Once a large body of information about a species is available, it is tempting to assume that this information applies to all other species, or at least all other related species, so that all types of deer are expected to behave like red deer, and all flies like *Drosophila*. There are at least two problems with this assumption. First, how far can we generalize? Is information about red deer applicable not only to other deer, but to sheep, or all grazing animals, or all land mammals?

The second problem is more serious. If we use model systems as the archetype, the quintessential example of their kind, it is easy to conclude that anything that deviates from the model is aberrant, not "normal," and that is if we see the deviation at all. My own research, which includes looking at mating behavior and the evolution of sexual characteristics, has made me increasingly aware of the ways our biases about gender influence how we view animal behavior. We tend to use males as a model system, not just in some animals, but in life as a whole, and that skews our perspective on both sexes. And that is why our friend Wing Bamboo called the loon "Brother."

THE "OTHER" WOMAN

Although attitudes are finally changing, the paradigm in science has been to view the male of a species, including humans, as the norm, and females or women as variations, special cases, exceptions to the rule. Analyzing this way of seeing things, Simone de Beauvoir called women the second sex in her 1949 book with that title. Psychologist Carol Tavris attributes the viewpoint to the "mismeasure of woman," in her book by the same name. Anthropologist Sarah Blaffer Hrdy wrote about "the woman that never evolved." I first noticed it in a rather personal way when I was a graduate student at the University of Michigan, which has a large medical school. Posted around campus were ads for volunteers to participate in medical research, the kind where the subject was injected with some potential allergen, or required to exercise and get blood taken at regular intervals, or some such procedure. The subjects were paid some sum like $25 a session, which at the time was a lot of money to me, so I was eager to participate. But whether the research was on exercise physiology or kidney function, the ads almost always called for men. I was taking a women-in-science seminar at the time and became intrigued by the idea that people used males as the model system. For my report to the seminar, I systematically surveyed papers in several physiology journals for the sex of their subjects, whether these were rats or hamsters or monkeys—deliberately choosing areas such as pulmonary function and circulation, where the reproductive systems weren't the topic of study. It turned out that scientists overwhelmingly chose males as their subjects. I also checked to see whether male rats were easier or cheaper to obtain, and it turned out they were not. This bias occurs in medical texts, in research, and in tests of drug effectiveness. The rather disturbing result is that virtually all discussions and diagrams of normal functioning of the liver, kidney, respiratory system, and most other nonreproductive aspects of the human body use the proverbial "70 kilogram man" as their subject.

What I was often told was that somehow the reproductive cycles of the females would obscure the "real" findings, as if the kidney was somehow a sexless structure able to be examined in isolation. The cyclical nature of female reproduction in mammals was seen as unwanted "noise" obscuring a view of the "real" system. This seemed peculiar to me for two reasons. For one, the reproductive cycle adds no more "noise" than the workings of the eye add to a study of other sense organs; no one would suggest that hearing is usually best studied in the blind, even if the simultaneous operation of auditory and visual systems creates interactions between the two. This is not to suggest that audition or any other physiological function would not bear examination under a variety of circumstances. The point is that female variation is no less valid than any other form of human variation. The second problem is that even if women's reproductive cycles do influence drug effects, for example, this seems like very valuable information, if we have any hope of treating women, reproductive cycles and all, using the drugs whose effects we are studying. Current research practices are not so biased, and a recent directive from the National Institutes of Health calls for more research into the biology of women, but the basic notion is still with us.

Perhaps the best illustration of this common perception of female as "other" comes from a study on mental health in humans conducted by I. K. Broverman and colleagues over a quarter-century ago. Participants were mental health clinicians given a sex-role stereotype questionnaire consisting of 122 dichotomous items such as "not competitive . . . very competitive" and asked to describe a "healthy, mature, socially competent (a) adult, sex unspecified; (b) a man; or (c) a woman." The results were, to my mind at least, distressing. Males and females were described very differently: the description the subjects gave of a man was virtually identical to that of an adult, while women were seen as distinct from adults. Inescapably, then, women cannot be both feminine and human, because the model system is male and females cannot be its representative.

Perhaps to our relief as social observers, this study has been criticized by sociologists as having painted an exaggeratedly biased picture of sex role stereotypes. But its main point and relevance to other aspects of science, including my field of behavior and evolution, remain clear: scientists, like other people, tend to view what males do as the norm, the model system, and what females do as a variation on the theme, a subcategory. Even if this attitude does not make females less studied to begin with, it can make them seem less interesting, or like a special case that should be dealt with only after the important individuals have been described or tested.

A good recent example of the male model in biology comes from a series of monographs called *Birds of North America*, which is an exhaustive survey of each and every species of bird on the continent. The account of each species was written by an expert or team of experts, and each one has a color photograph of the bird on the front cover. Virtually all the photos feature a male as the species representative, but when the feminist scientist Patty Gowaty was invited to write about "her" species, the eastern bluebird *(Sialia sialis)*, she rebelled. She wrote the editor of the series, Alan Poole, and asked for special dispensation. As she described it to me, she wanted:

> the editor to please, please just this time to put a photograph of a male *and* a female. "After all, I do study monogamy, it would be appropriate, besides females are pretty," I begged. Stony silence. I brought it up at least 8 times. Alan was really a stone. When the box of reprints came, I peeked in, and with great disappointment exclaimed

"There's only one bird in the photograph." Then I looked at it. The front has a single gorgeous *female* bluebird. There is no legend. I actually do consider this my most significant accomplishment in "the feminist construction of science."

Some of the best examples of the use of male model systems in behavior come from primatology, the study of monkeys and apes. Sarah Blaffer Hrdy, a pioneer of the female perspective in the study of primates, said that with regard to females, "amazing as it sounds, only relatively recently have primatologists begun to examine behaviors other than direct mother-infant interactions that affect the fates of infants." She was referring to the last twenty-five years of primatology. Female monkeys were and are often equated with mother monkeys, as if no other role could exist. Similarly, my students in a field biology class almost always call animals they see "he." Birds, mammals, reptiles, and insects are all "he," and it is difficult to train oneself out of this generalization even when it should be obvious that males and females may behave differently in the field, making pre-judgments about the sex of an animal risky.

In the case of social insects such as ants or bees, one can actually miss the pertinent biology, because in most species only females forage for the colony or perform other tasks best not interpreted in light of their repro-ductive roles. All the bees one sees flitting from flower to flower are females, workers produced by the female reproductive, or queen, who remains in the hive. Because of their production of honey and their usefulness as pollinators, people have taken an active interest in bee life for thousands of years, with ancient Egyptian depictions of beekeepers dating back to 2400 B.C. It turns out, however, that while for much of this time everyone had recognized the hierarchy of bee society, they had also thought the colony was governed by a king, a male. It was not until 1637 that Jan Swammerdam demonstrated by dissection that the ruler was indeed a female. (Note, incidentally, that microscopic examination of the insects was hardly necessary to draw the correct conclusion; one could simply have observed which individual laid eggs.) It is irritating if not surprising to see that in two recent children's films worker ants were portrayed as male. But stereotypes are hard to break. And so we have Wing Bamboo, politically correct though he may have fancied himself, viewing the loon as his brother, and not some female relative.

Now, some might argue that this is "just" terminology and that it is not the major issue. The importance of language in structuring our thinking, and the use of "man" as a false generic, are beyond the scope of this chapter, but it is clear that how we speak often unconsciously constrains what we think. Children asked to draw a fireman draw a male, while those asked to draw a firefighter may draw one of either sex. (When asked to draw a scientist, most draw a man as well; the children also think he is white and more likely than not has facial hair and glasses.) Furthermore, if females are described only as mothers, we are less likely to derive theory about, say, foraging behavior, or learning, using females as models. The model system becomes less of a simplifying guideline and more of a role model, something to emulate. If we always see males as the norm, a society in which males are dominant also becomes normal. We may miss what females do in our own culture and in those of animals.

Many people in recent years have seen these biases and objected to them. Perhaps in reaction to such restricted viewpoints about females, more recent work on animal behavior promotes the discovery that many supposedly feminist relationships and behaviors are much more common among animals than had previously been believed. It turns out, for example, that sexually aggressive females and sexual behavior outside a pair bond are common, which means that sex roles are not defined as narrowly as some would have us think. Males can be excellent caregivers, and females want sex as much as males. We can take our inspiration from bonobos, which show female-female sexual interactions. Or we can look to several species of butterflies, in which females mate with several males in succession and seem to actively manage the sperm of each of them when fertilizing their eggs, instead of to the domineering male baboons that herd females with the threat of violence.

I greet these discoveries, and their emphasis in the media, with mixed emotions. Is this really where we want to go? Do we want an escalating argument in which opponents cite examples that support their ideologies? We could have an endless debate in which one side points out that male elephant seals may crush pups (conclusion: females are at the mercy of the larger, more powerful sex), to be countered by the finding that female bonobos are sexual in a variety of contexts (conclusion: female sexuality is a flexible behavior). Next round: with few exceptions, males are physically larger and behaviorally dominant, at least in many vertebrates, followed by the "yes, but" statement that in many animal social interactions, size, at least body size, doesn't matter. Presumably the feminists could triumphantly put an end to the discussion by pointing out that female praying mantids often consume their mates.

Ultimately, though, what is the point of these arguments? Using animals to inspire or support our ideas about social justice is bound to fail. It also has a great danger of backfiring; what if, for example, we find that we were mistaken about female control of paternity, or female sexual aggressiveness? Does that mean that women should attempt to be passive and subservient? One hopes not. I like to think we are perfectly capable of choosing our own visions of an ideal society without the help of snow geese or gorillas. We do not need a new model system, even one with a more pluralistic bent. Yes, recognizing diversity is good. We do want to make sure the actions of females are not ignored. More important, however, we need to stop confusing model systems with role models.

In a classic work on nature, *The Outermost House* (1928), Henry Beston wrote a passage about animals that has been embraced by environmentalists:

> We need another and a wiser and perhaps a more mystical concept of animals. Remote from universal nature, and living by complicated artifice, man in civilization surveys the creature through the glass of his knowledge and sees thereby a feather magnified and the whole image in distortion. We patronize them for their incompleteness, for their tragic fate of having taken form so far below ourselves. And therein we err, and greatly err. For the animal shall not be measured by man. In a world older and more complete than ours they move finished and complete, gifted with extensions of the senses we have lost or never attained, living by voices we shall never hear. They are not brethren, they are not underlings; they are other nations, caught with ourselves in the net of life and time, fellow prisoners of the splendour and travail of the earth (p. 25).

The environmentalists are fond of the phrase that cautions us not to see animals as underlings, because conserving nature means that we stop seeing other species as objects only here to be used for our own purposes. As a scientist who studies behavior, I think the view of animals as brethren is at least as dangerous, and it is dangerous in both of the ways illustrated by Wing Bamboo and the dead loon. The loon is not a brother, or a relation in the sense that many of us would like, and furthermore it is risky to assign it a gender without knowing it. The way we see animals is central to how we treat other human beings, the earth, and of course the animals themselves. Ultimately, understanding animal sexual behavior will show us an undreamed-of diversity. Even more important, it may humble us in our assumptions about what is natural, normal, and even possible.

Two

SUBSTITUTE STEREOTYPES

The Myth of the Ecofeminist Animal

DISCOVERIES IN BEHAVIORAL ecology and evolutionary biology that inform us about what the sexes are like can potentially change our ideas about what it means to be male and female. Information about what particular animals really do exposes some stereotypes that, I have argued, scientists' own biases about the role of females in general and women in particular have sometimes prevented us from recognizing. A feminist viewpoint that is aware of those biases can be very valuable in understanding animal (and human) behavior. But there is a danger in rejecting stereotypes, namely, the risk of simply substituting a new, perhaps more politically correct one. This may serve the politics side of the debate, but ultimately it badly lets down the science side. In this chapter I explore how trying to use science to further a feminist agenda is equally damaging to our efforts to understand the natural world.

Links between feminism and the study of nature have been made many times by many different people, including some of the early women naturalists, who felt their feminine outlook gave them a unique ability to appreciate the plants and animals around them. Perhaps the most recent and strongest of such links comes from the ecofeminist movement, which draws a connection between ending inequality between the sexes and solving environmental problems by changing the human relationship to the natural world. Three different angles to ecofeminism make this association, and writers associated with various strands of the movement have pointed, with different degrees of emphasis, to all three. First is the idea that women have more "connections to the primal," as Mary Morse puts it, or that "women have a greater appreciation of humanity's relationship to the natural world, its embeddedness and embodiedness, through their own embodiment as female," as Mary Mellor suggests. Because of their reproductive and care-giving functions, women are often traditionally seen as less cerebral and more physical or "natural." Instead of rejecting this viewpoint by striving to become detached and analytical, ecofeminism embraces the supposed emotionality and spiritual connections of women and uses them to attain a deeper and presumably better understanding of the earth. A related, and older, suggestion is that women are naturally more peace-loving and less aggressive than men, so they are better suited for making this a more compassionate world. These views are also similar to those of the so-called difference feminists, who believe that men and women are inherently different but that the problem in society is that female qualities, such as intuition or emotion, have traditionally been denigrated in favor of more male ones, such as objectivity.

The second link is between abuse of the environment and abuse of women; the reasoning here goes that because these stem from a similar source, namely the culture of men in power that subordinates nature and victimizes women, solving one side of the problem requires solving the other. Women experience firsthand the effects of environmental degradation through miscarriage, birth defects, and contaminated breast milk, and hence may be more sensitive to the need to eliminate the problems and more motivated to do so. In *Feminism and Ecology*, author Mary Mellor quotes Maria Mies, who says, "in defying this patriarchy we are loyal to future generations and to life and this planet itself. We have a deep and particular understanding of this both through our natures and our experience as women [such that] wherever women acted against ecological destruction or/and the threat of atomic annihilation, they immediately became aware of the connection between patriarchal violence against women, other people and nature" (p. 105). On a more practical level, it makes sense to be concerned about the quality of the environment women are fighting to be equal in; Ynestra King observes, "The piece of the pie that women have only begun to sample as a result of the feminist movement is rotten and carcinogenic, and surely our feminist theory and politics must take account of this, however much we yearn for the opportunities that have been denied to us. What is the point of partaking equally in a system that is killing us all?" (p. 106). Perhaps the most recent manifestation of this view is the involvement of ecofeminists in the controversy over genetically engineered crops or, as they are sometimes called, Frankenfoods.

Third, several writers have gone further, and concluded that science as a way of understanding the world is fundamentally flawed because it is founded in antifemale, antinature attitudes that emphasize domination and objectivity at the expense of harmony. Carolyn Merchant, Evelyn Fox Keller, and Mary Mellor all hold to varying degrees the viewpoint that Western science in effect "killed" a Nature that was seen as both female and alive, forcing a separation from the living world and from the essence of being female. The scientific revolution of the sixteenth and seventeenth centuries thus paved the way for what Mellor calls "a disenchantment of Nature," a turn of phrase I find curious and to which I will return shortly. This aspect of ecofeminism also leads into the suggestion that science itself is bad for women, both because it champions this detached, unfemale perspective and because in it lie the seeds of destructive biotechnology. Therefore we should strive for a more holistic understanding of nature by using female qualities to inform us.

As both a feminist and someone who has championed environmental causes for most of my life, I was predisposed to like the merging of the two. It initially seemed a good idea to link eliminating oppression of women with eliminating misuse of the earth. I found, however, that I became disturbed where I would have liked to be heartened, alienated where I would have expected to find sense, and at worst put off by the sheer sentimental twaddle that seemed to be espoused in the name of humanizing our understanding of the natural world. What's wrong with this new picture of the female embedded in nature, and why doesn't it seem to teach us very much about the animals with which we have supposedly become better connected?

BEING DIFFERENT DOESN'T MAKE YOU WORSE, BUT IT DOESN'T MAKE YOU BETTER

The premise that females are the more compassionate or gentler or more nurturing sex is a shaky one on several counts, and I hasten to point out that many other feminists are also uneasy about invoking the idea that women can save the planet simply by virtue of being women, or that if women are not inferior to men they must be superior instead. Along with many other scientists, I am particularly suspicious of the notion that females in general have some sort of caring nature that makes us less aggressive and enables us to empathize with other organisms better. Presumably, if this compassion is part of our nature, we share it with other animals—where else, after all, could it have come from? So let us first look at other species and see if this assumption is well founded.

In the next chapter I examine maternal instinct and the idea of the nurturing mother, and find them to be far from universal in the animal world. What about other attributes of females, particularly with regard to aggressive behavior? Many people believe that if women ran the world, wars would end, gun control would be unnecessary, and we would all cooperate to achieve common goals for the good of society. This may or may not be true; what certainly isn't true is that whatever pacifist inclinations women possess come from our "nature," which we share with other female animals. Unfortunately for these and other utopian notions, it turns out that it's a warbler-eat-warbler world out there. Literally, in fact.

The great reed warbler (*Acrocephalus arundinaceus*) is a bird that migrates each spring from Africa to Europe, where it settles in reed beds of lakes. Great reed warbler males may attract more than one female to settle on their territories. When they do, the males differentially allocate their help with the offspring depending on when the female has settled; the first, or primary, female gets a greater share of fatherly aid in the form of fetching insects for the babies than the secondary or later females. So life is good for primary females, unless they lose their clutch, in which case the male turns his attentions to any other females on his territory. Then Swedish ecologists Bengt Hansson, Staffan Bensch, and Dennis Hasselquist noticed that a suspiciously large number of eggs of the primary females were disappearing under mysterious circumstances. It did not seem to be random loss from predators, because the primary females were three times more likely to lose their eggs than secondary females, and predators such as crows should not have any reason to target the nests of the primary females in particular.

So the scientists performed an ingenious experiment: they placed plasticine eggs in artificial nests on the great reed warbler territories and compared the peck marks left in the claylike surface with marks made by known species of birds and mammals. Predators always test an egg by biting or pecking it before they remove it from the nest, and if the object turns out to look like an egg but not feel like one, they often go no further and leave the little oddity alone. The nibble made by the bill of the warbler is quite distinctive, and Hansson and his colleagues were able to deduce that the eggs of the primary females were being destroyed by secondary females on the same territory, a behavior that increased the likelihood of garnering a higher proportion of male parental care for themselves and their own chicks. This infanticide is hardly an example of pacifism among females, but scientists, while intrigued by the results, were not surprised by them. Note, too, that it does not require invoking conscious malice on the part of the warblers; females that responded to the presence of eggs by removing them when they arrived on a territory did better in terms of their own reproduction than more oblivious females, and thus natural selection could have caused an increase in such responsive individuals.

Many other examples of vigorous female competition and aggression have long been noted in birds and other vertebrates. Even that classic symbol of cheer and optimism, the bluebird, turns out to have a dark side. Patty Gowaty, whom I have mentioned in the previous chapter, found that female eastern bluebirds *(Sialia sialis)* were so aggressive that they sometimes even fought to the death. Bluebirds are what ornithologists call cavity-nesters; instead of setting their nests in the branches of a shrub or tree, they build them in holes that form in dead tree stumps (or in boxes placed out in fields by investigators). Under natural circumstances, these holes do not form very often, and thus nest sites are quite limited and females battle fiercely over them. Gowaty thought her results, while interesting, were not particularly out of line with what one would predict, given the scarcity of nest cavities for females and the potential for big wins in terms of reproductive success if other females are kept out of the territory. She was therefore taken aback when her discovery was deemed so sensational that it made it into the "Ripley's Believe It or Not" cartoon in the newspaper, a status more usually reserved for two-headed calves and people with 500-pound balls of string. Presumably it seemed extraordinary because, once again, reality did not conform to the stereotype. Killer bluebirds are bad enough, but female killer bluebirds seemed to be the stuff of nightmares.

The real champions of female brutality, however, come from the species where females rule, and in a monarchy so absolute that insubordination is seldom tolerated. I am talking about the social insects, among which females comprise the vast majority of individuals in the colony and males play a small though crucial role by fertilizing the next generation of queens. Severe competition among the females has been documented since the 1960s, and biologists such as Mary Jane West-Eberhard in Costa Rica have studied it intensively. The strife appears at several levels; in species like the honeybee, the first queen to emerge from her pupal cell may kill the others as they lie in their virginal waxy chambers, or the female workers may indirectly rebel against the queen's reproductive tyranny by manipulating the number of males or females they rear in the hive. The workers may also subversively slip a few eggs of their own into the hive, if their reproductive organs can develop sufficiently.

In many species of wasps, females battle fiercely for superiority when a colony is founded. Wasps form colonies of up to a few dozen females, rather than hundreds or thousands of individuals like the bees. They may build nests from mud, like the mud daubers, or from chewed-up wood pulp, like the paper wasps. Unlike the honeybees, many wasp species show relatively little differentiation among individuals, so that instead of a morphologically distinct class of sterile workers and a few queens capable of laying eggs, most or all wasp females are physiologically able to produce offspring. The problem is that females do better in groups, because they can fend off predators more effectively, but within that group, being the dominant individual pays off much better in terms of reproduction than being a member of the rank and file. So females keep together, but each one strives for the top position. Vicious fights result, with limbs lost, antennae torn from their sockets, wings raked by mandibles. These battles are among the most violent in the animal kingdom.

Life among the social insects has always seemed rather ironic to me in light of the nostalgia among many for the lost days of Goddess worship and matriarchy. According to a much-popularized view of our early history, humans used to live in benevolent societies that worshipped female deities, venerated female qualities, and were closely in tune with the natural world. Only when male-dominated, Judeo-Christian philosophy took over did humanity begin to be divorced from the earth. Carol Tavris, in her marvelous book *The Mismeasure of Woman,* calls this belief "the search for a feminist Eden," and expresses considerable skepticism about a time when "religions were peace-loving and woman-worshipping, deities were female, women were priestesses, and neither sex wore the pants, as it were" (p. 72). This all gets involved with another feminist issue, body image, since some of the support for such a scenario comes from comfortingly plump female figurines that supposedly represent the good old days before we had to worry about cellulite and tight jeans.

As Tavris and many others have argued, not only does considerable doubt exist about the reality of such an egalitarian past; even if it had been so, it is not at all certain that this history should provide the impetus for changing our current society. My own point with the wasps is that one need not search for a primordial Nature with a female-dominated society;

we already have it, and I for one do not want to live there. It is the perfect case of needing to be careful what you wish for. A land of milk and honey—at least the honey part—seems to come at the expense of having your head ripped off by another female.

Of course, we hardly have to model a female-friendly world after paper wasps. But that is precisely my point: looking for support for a nature of cooperative, loving females is just as foolish as looking for support for an animal equivalent of Ozzie and Harriet. It is good to debunk the myth of the passive female, good to look for positive images of women in nature and art, and it has been surprisingly hard to do so, a testimony to prejudices of both the scientists and the public. But substituting one stereotype for another will only defeat our goal, and what is worse, it will still prevent us from learning about what the animals really do. There is no reason to expect that a female-oriented society will be a utopia, and there are some pretty compelling ones to expect that it would not.

DUALISM AND THE NOT-SO-OPPOSITE SEX

As numerous psychologists have noted, people like to divide the world into opposites (like the old joke about there being two kinds of people, those who think there are two kinds of people and those who do not). Whether it is because there happen to be exactly two genetic sexes, or whether having two sexes simply provides a convenient example of this dualist thinking, nowhere is such a litany of us/them as pervasive as in comparisons of what males vs. females are "like." Males are aggressive, females are passive. Males are independent loners, females are interactive connection-builders (or gossipy, if you prefer). Males are exploratory, females are nest-builders and care-givers. I could, of course, go on.

There are two major problems with this categorization. (Oops—I did it myself. Make that at least two.) The first one has received a great deal of attention, and is the difficulty highlighted by the Broverman study of mental health in men and women that I mentioned in the previous chapter: male attributes are often the ones we favor, and the ones we think are necessary for success in the world. This division has been bad for women, because of course being identified with characteristics that no one holds in high esteem just perpetuates the treatment of females as inferior. As a result, it is tempting to counter either by claiming that women, too, are warriors and explorers just like men, or by celebrating the female qualities but keeping the dualism intact. Thus we have the ecofeminist model of the goddess-woman, who doesn't deny her earthy nature but rejoices in it.

Either way, though, we run into a different problem: the dualism itself is flawed to begin with. Males are not always this way and females that way, at least not in any immutable sense that makes it impossible for us to seek change. Yes, selection has acted differently on males and females of all animal—and for that matter, plant and many microorganism—species. The behaviors that benefit your average female wasp are different from those that benefit the average male wasp, or bluebird, or pipefish. I like many other scientists have made a living out of exploring those differences, and I have no doubt that they are there, manifested in different ways under different circumstances. This action of selection often means that the sexes are in conflict, and that males and females compete not only among themselves but with each other. But it does not mean that the differences are fixed, with all creatures possessing the same extreme values for masculine and feminine characteristics. Even if males are a certain way, it does not mean that females must be their opposite. And it certainly does not mean that if we find differences, we have to set them in a hierarchy, so that one is better, whether that one is the male or the female way of doing things. For whatever reason, categories often seem to mean ranks, so that being able to describe something means rating its merits relative to something else. If males are aggressive and females passive, we seem to want to see either aggression as good or aggression as bad, and the male way or the female way has to prevail.

But as both human beings and scientists we are better off without this kind of ranking. We are better off without having to champion killer bluebird females as being more politically correct than stay-at-home bluebird females simply because the feisty ones suit our image of independent, in-your-face women. We certainly need to debunk the stereotype of the passive female, but we need to do it because it is wrong, because it can prevent us from seeing the female bluebirds fight, not because aggressive female bluebirds are more our style. The bluebirds do what the bluebirds do (I will take as given that we can in fact observe them doing something; in other words, bluebird behavior, like the behavior of atomic particles or blood cells, occurs in the real world and is not just in our heads, as some of the postmodernists would have us believe). Part of my job as a scientist is to clear my preconceptions out of the way as much as possible so that I can see what the bluebirds are doing.

At the risk of taking on the question of whether science can ever truly be objective, I now want to consider the ecofeminist claim that science itself is antifeminist because it places human beings outside of nature rather than considering us a part of it. The question of whether science has separated us too much from the natural world is an old one, and one that touches on our relationships with other organisms as well as the way in which we think we "understand" something. Mary Morse flatly states that science itself "sanctions domination of both nature and women," a view echoed by other ecofeminists such as Carolyn Merchant. Evelyn Fox Keller draws parallels between the controlling nature of scientists and mental diseases such as paranoia and obsessive-compulsive disorder. While hardly consigning all of us to the psychiatrist's couch or the mental institution, and still acknowledging that we scientists differ in our approaches, she does "suggest that a science that promises power and the exercise of domination over nature selects for those individuals for whom power and control are central concerns. And a science that conceives of the pursuit of knowledge as an adversarial process selects for those who tend to feel themselves in adversarial relation to their natural environment" (p. 124). Or to use the trendier term, we scientists are all control freaks.

As a scientist, and particularly as one who studies animal behavior, I find almost touching the faith that nonscientists have in my ability to control and predict natural phenomena. Let me tell you a secret: it's a mess out there, with "out there" being the natural world we are all supposed to be trying to keep under our thumbs. The suggestion that we control nature contains the implicit assumption that we already understand it, an assumption that becomes more and more wildly incorrect the longer we work on natural systems, both biological and physical. It is not that we never learn anything, it is just that each small piece of the puzzle not only helps answer the current question, it invariably suggests other puzzles, each with thousands of pieces, lying in their own boxes underneath the one we are working on at the moment. For me, doing science and studying nature have made the world seem less controllable, not more. Most scientists that I know find the world more complex than the nonscientists do. If you do not really look at something, it is easy to oversimplify it.

This belief in the ability of scientists to control the world was first brought home to me by one who was a scientist himself, trained as a psychiatrist but working in the field of evolutionary biology on human behavior. While still a graduate student, I gave my first talk on my research at a small conference. It concerned the mating behavior of two species of field crickets that are common in lawns and fields of the eastern half of North America, a topic that became my doctoral dissertation. I was particularly interested in how a one-celled gut parasite affected the males' ability to attract females for mating. After it was over, the psychiatrist came up to me. "That was a good talk," he said. "It must be nice to work on a species where you really understand everything about it." I was stunned. I looked at him closely to make sure he was not being sarcastic. Surely he did not mean that I or anyone else understood everything about what crickets did, or everything about the dynamics of the parasite in the crickets, or even everything about how the parasite influenced male behavior. I finally realized that from his perspective, as someone who studied the intricacies of human motivation and emotion, crickets must look like vastly simpler versions of people, chitinized humans without the big brains, complex social groups, or material goods. This is one version of the controlling scientist story: because we do not want to think about the web of human complication, either in society or literature, we try to reduce the world to basic elements in the form of animals which are crude caricatures of the "reality" of emotions and culture.

Nothing could be further from the truth. This mistaken attitude is another misuse of model systems, as if we really would rather study humans but study animals instead because they are less complex. Then it is a short step to conclude that we must be able to understand everything about the animals, and so can enter into a controlling relationship with nature. But we do not understand everything, and even if scientists have been able to explain how people might use nuclear power and release ladybugs to eat agricultural pests of crops, most scientists that I know find the world more complex than the nonscientists do. And not having the understanding may make us frustrated, but it does not need to make us adversarial.

This takes us back to Mellor's and others' ecofeminist view that "masculinist science" and "the scientific revolution brought about a disenchantment of 'Nature.' Wild and alive Nature was replaced by a mechanistic world view that saw the natural world as dead and passive. Whereas the organic view had restrained exploitation, or at least made it self-conscious, the mechanistic view, associated with Descartes and Newtonian mechanics, led to the 'death of nature', as an idea and in practice" (p. 116).

This attitude is profoundly strange to me. Why should close observation and asking questions about how something works rob it of its vibrancy?

Why should Nature have become disenchanted when we examined it? Is it some gigantic version of familiarity breeding contempt, so that understanding animals made them less appealing? Far from diminishing my appreciation of the natural world, studying organisms has done everything to increase it. Nature is much more enchanting to me now than it was when I knew less about it.

Women often complain that men do not listen to them; paying attention to someone is the greatest compliment you can give them. Scientists are people who pay attention to the natural world, and I do not see how one can truly interact with nature without paying this attention. This attentiveness may seem mechanistic to some, but the alternative is fuzzy and somewhat condescending, like a loving but uncomprehending parent patting a child on the head and murmuring, "There, there, everything will be all right," without understanding the child's problem or having any reason to think that it will, in fact, be all right.

PARTNERS AND STRANGERS

What is the alternative to scientific "domination"? Carolyn Merchant suggests a "partnership ethic that treats humans (including male partners and female partners) as equals in personal, household, and political relations and humans as equal partners with (rather than controlled-by or dominant-over) nonhuman nature" (p. 8). This sounds fine, but I would be hesitant to enter a partnership with someone I knew only slightly. Presumably the ecofeminists wouldn't share a household with a perfect stranger, or start a business or raise a child with one. What is this nature with which—or with whom—we become affianced? And what kind of partnership can we have with creatures we know little about, because studying them represents a controlling patriarchy? If we simply rely on feeling a connection with them, but don't know how they breathe or reproduce or find food, we have learned more about ourselves than about them.

The risk is that we will idealize nature, view it and its plants and animals as a new feminist version of the Noble Savage. A feminist perspective on the world should not mean putting animals on a pedestal in the name of being respectful of them. The danger then is requiring them to live up to your expectations. If nature is automatically pure and good, what to make of predation? What to make of lions and langurs that kill their young, of the cute little squirrel that is transformed into hawk flesh? If we are not supposed to dominate nature or each other, what of domination within nature?

However reluctantly, most of us recognize that wolves must eat too, and sometimes they must eat Bambi. Slitting the throat of a deer is at least reasonably quick. Some behavior may seem unpleasant, but it is sometimes a necessity. But it gets worse; what to do about parasitoids?

Parasitoids are part parasite, part predator: they live inside another animal, like a parasite, but they eventually kill it, like a predator. I still work on crickets, but lately some of my research has concerned one of these parasitoids, a fly that hears the male cricket calling to attract a female and homes in on him herself (only the female flies do this). Once having arrived, she deposits sticky little larvae, each only slightly larger than one of the commas in this sentence, on and around the male. In minutes, one or a few will bore a hole through the outer skeleton into the cricket's body and nestle in the tissue under the wings. They will spend a few days eating cricket insides and constructing a tube to breathe through, like a tiny periscope. In a diabolical twist, they use the cricket's own cells that circle in response to the invasion for making the tube, and in so doing seal the doom of their host. After the maggots have grown somewhat, they move to the fat in the cricket's abdomen, and continue to feed and grow until, after about a week, two or three individual larvae can occupy virtually the entire body cavity. From the outside, it looks like a cricket, walks like a cricket, and—until the end is quite near—even talks like a cricket, able to produce the deceptively cheerful chirps that signal summer. Except for this brittle shell, however, almost all the cricket meat has been converted into fly, slowly, while the cricket is still alive. Finally, the larva has grown enough, and it shatters a hole in the side of the host, bursts through, and digs into the soil, where it pupates and eventually emerges as another adult fly. Only then, when it is no longer useful to its parasite, does the cricket host die.

I have dissected hundreds, perhaps thousands, of crickets over the last several years, and when I discover one of the maggots, white and pulsating, it is always with shock and amazement. I tell audiences when I discuss this research that everyone in the world, scientists and nonscientists alike, should open up an insect and see a parasitoid at least once in their lives, and the audiences laugh, but I am not entirely joking. It is worth seeing something that horrific because it forces you to reexamine that warm fuzzy feeling about nature. I am not suggesting that instead we eschew what animals are, and make up some equally self-serving story about cruelty and the law of the jungle. Nature, remember, is witless. It is not inherently evil any more than it is inherently good. I just want to make sure that while we discuss our relationship with nature, as feminists or anything else, we actually talk about nature itself and not simply about the relationship.

I also do not mean to suggest that there are no ethical issues in doing science, or that all ways of studying animals are justified without regard for the pain and suffering that the animals may feel. I have the utmost respect for those people working in animal welfare, trying to find a way to study animals without exploiting them. What kinds of experiments or manipulations are acceptable? Is depriving a monkey of her mother at birth a legitimate research technique? What about depriving a dung beetle, which also experiences a long period of parental care? Why are we more likely to veto the first but not the second? What about experiments that reduce food levels for animals? All right for a short time, not all right for life? Answers to these questions are beyond the scope of this chapter, but it is undeniably important to consider animals from an ethical viewpoint, and most of the scientists I know do so.

I do not want to study animals only to learn about me, though that may happen along the way. I want to learn about the maggots. This may involve the kind of detachment that the ecofeminists decry. But what I advocate is not so much detachment or domination as an ability to remove oneself from the center of things. I know that my biases will influence my interpretations, but at least we can struggle to see what animals are doing. This may be patriarchal science, but I do not think so.

In an ecofeminist book called *Reweaving the World*, Brian Swimme writes, "My own hope is that what is happening in our time . . . is the emergence of the common myth necessary for us to feel and act as kin to everything" (p. 22). This is well and good, except that it takes us back to Wing Bamboo and his kin the loon. What does it mean to feel and act as kin? At best it is vague and soppy, at worst misleading. It tells us something, perhaps, about Wing Bamboo. But it says nothing about the loon.

Three

Now THAT WE ARE ON GUARD against various sorts of stereotyping, and have dispensed with the need to use other animals as role models for our own behavior, let us move on to a consideration of one of the most basic characteristics of female organisms: motherhood. This is a sensitive topic, entwined with our notions of selfishness, sacrifice, and femininity, and one in which misconceptions abound. One of those misconceptions is that caring for one's offspring is synonymous with being female, a view that has constricted our ideas both about what else females do and about how offspring care occurs. Ironically, it turns out that many of biology's most profound mysteries center around what is essentially child care, the less than central "women's issue" of politicians. It is the place where extreme selfishness becomes the same as extreme sacrifice.

THE MATERNAL INSTINCT

Let us first examine the idea that motherhood and maternal care are "natural," the one type of behavior that females know instinctively how to perform. If maternal caregiving behavior is natural, if it is "biological" (by which term the popular media usually sidestep the question of a genetic basis to a behavior), then presumably all women want to do it, know how to do it without learning how, and feel deprived if they do not. A modern version of the maternal instinct is the mother-infant bond, this mysterious connection that will naturally occur between a woman and her child.

47

How fixed is this bond, and how ingrained in a female's psyche is the ability—and even the desire—to care for her young? For a start at an answer, it is instructive to look at some research done in the 1950s and 1960s with rhesus macaques, the monkeys used in numerous medical and behavioral studies; the "Rh" in Rh-positive or Rh-negative blood types comes from the name of this species. Two psychologists named Harry and Margaret Harlow were interested in understanding the effects of social deprivation on humans, the kind of deprivation that arises through abuse, neglect, and warfare. Children in orphanages in wartorn countries, for example, show many aberrant behaviors. The Harlows wanted to know how these problems arose and thus perhaps gain insight into correcting them. They could not, of course, perform controlled experiments on human children, and so they did what seemed like the next best thing: they manipulated the early rearing environment of groups of macaques so that deprived youngsters could be compared to more normally raised controls.

The infant macaques were taken from their mothers at birth, and raised under one of several different sets of conditions. For example, some were allowed access to a monkey-sized wire model that had a bottle attached to it approximately where the nipple of a mother monkey would be. The bottle was filled with milk and the cages cleaned at appropriate intervals by human caretakers who had no contact with the monkeys. Other baby macaques had the wire model with the food, and in addition were given another model covered with terrycloth. In some instances, the monkeys saw or contacted other equally deprived infants during their development, while in others they were kept in isolation. As the monkeys grew up, they were used in a variety of tests to examine their social behavior and eventually their ability to mate and have offspring of their own.

To the surprise of few, the socially deprived monkeys did not exhibit normal behavior. They were more fearful when new stimuli were introduced to their cages, they did not interact with other juvenile monkeys the way that babies raised with their mothers did, and they showed numerous other signs of psychological abnormality. Interestingly, however, the babies given the cloth-covered model "mother" were better adjusted than those given only the wire model, and when a scary object like a mechanical toy monkey playing a tin drum was introduced into the cage (I have always found this choice of stimulus to be a bit too ironic), they ran not to the source of food but to the source of what the Harlows called "contact comfort." In other words, clinging to a soft object is more soothing than returning to the place where the essential milk of life is found.

The Harlows went on from these studies to discuss a number of fascinating theories about the need for this contact comfort and its potential value in helping children with minimal resources develop more normally. Their ideas have been thoroughly dissected—and in some cases debunked—by developmental psychologists. For our purposes here, however, a more relevant finding was the discovery that females raised with either the cloth model, the wire model, or both had difficulties in mating with male monkeys. They simply did not know how to have sex—they didn't know the postures, the signals, the responses. Even more significant, if they were artificially inseminated and became pregnant, they were incapable of caring for the resulting young, and the infants had to be removed from their mothers lest they be seriously injured. The mothers did get better with experience, so that subsequent young fared better, but the early babies were as foreign as extraterrestrials to their mothers.

In my opinion these results put paid to the notion of a simple maternal instinct. Even mother monkeys, mere animals acting on instinct, do not automatically know that they should be bonded to their infants, much less how to take care of them. The mysterious mother-child relationship is shattered when the mother is raised without others of her kind.

Well, you might say, this is clearly an unusual situation. Monkeys are not usually raised by wire models, and under natural circumstances they can relate to their offspring perfectly well. This, however, is precisely my point: even a behavior supposedly as sacrosanct as the love a mother will have for her child depends on the environment. Here, then, is as good a place as any to discuss the nature/nurture problem that keeps dogging our attempts to look at behavior and biology, by examining the nature of nurture itself.

What is an instinct? Without getting into a lot of tortuous terminology and the opinions of ethologists and other professional investigators, most people would agree that a behavior is instinctive if it develops spontaneously without a need for the animal to learn it. Instinctive behavior, behavior occurring in animals that appear not to have had any opportunity to see others do anything similar, has been noticed for thousands of years; Descartes discussed instinctive behavior. This definition would be okay, and reasonably workable, except for the baggage it carries: for some reason an exaggerated importance is often placed on what are termed instincts, as if we were powerless in their grip. Boys will be boys. We fear of course that if we acknowledge an instinctive basis to any behavior, it automatically becomes biologically determined, inevitable, which means that we had better keep the whole issue swept firmly under the carpet and focus on how television is the source of socialization evils and teaches us our sex roles.

It may surprise nonscientists to know that most behavioral biologists have long since abandoned the dichotomy of learned vs. innate, acquired vs. genetic, and nature vs. nurture. The reason is simple. Anyone who studies behavior quickly realizes that it is impossible to separate the environment from the organism experiencing it, so that all traits are necessarily the result of an interaction between the animal and its perceived world. This is not to say that one cannot attribute genetic and learned components to behavior; the entire field of behavioral genetics exists to explore the inherited basis of what animals do. The key point is this: only a difference between traits, and not a trait as such, can be said to be inherited or learned.

This is an old idea, and one that I learned most clearly from a book published in 1971 by Hans Kummer on primate social behavior. In it, he uses the analogy of human languages. The trait of "speaking French" is not classifiable as genetic or learned, because it contains elements of both. One cannot teach nonhumans French, and even people with certain genetic defects are not able to learn to speak. Hence, there is clearly a genetic basis to the behavior, else we would all be conversing happily away with our pets. At the same time, of course, though Francophones may deny it, speaking that language is not inherent in our (or their) genome, and though one can learn that "chat" means a small furry animal that purrs, one can just as easily learn that the same animal is called "gato" in Spanish. It isn't speaking French that is learned, it is speaking French rather than Spanish, or Dutch, or Urdu. So identical twins with the same genetic makeup, one raised in Paris and one in Amsterdam, would grow up speaking different languages, because speaking one language rather than the other is the aspect of the trait that is learned. Similarly, if one could subject genetically dissimilar individuals to the exact same environment, and they still emerged speaking different languages, it could be said that speaking one tongue rather than another is genetic. The difference between traits, not the trait itself, is the crucial element.

Kummer also points out that unless one were a hopeless pedant, one would never spend time contemplating whether a particular piece of music, say a violin sonata, was produced by the violin or the person playing it. One cannot separate the sound from the two vital parts that comprise it. It would, however, be perfectly legitimate to discuss whether two renditions of the sonata differed because they were played by different violinists or because they were played on different instruments.

I find this viewpoint tremendously liberating, and in various manifestations it is generally accepted by biologists. Russell Gray, a biologist from New Zealand, champions a somewhat different version in the form of developmental systems theory, but the point is the same: arguing over whether a trait, be it a tendency toward nurturing or a talent for mathematics, is genetic *or* learned is fruitless. All traits—even traits that seem blatantly one or the other—are both. The maternal instinct, as a behavior that arises absolute and predetermined from its primordial genetic roots, is a myth.

FEMALES ARE NOT THE EQUIVALENT OF MOTHERS

If one is freed from the idea of the maternal instinct, the belief that mothering is necessarily "natural," one is also freed from equating being female with being a mother, as if no other role was possible or important. Sarah Hrdy, the well-known primatologist and anthropologist who brought to our attention infanticide in a kind of monkey called a langur (more on this rather unparental behavior later), was the first to articulate the limitations of equating female primates with mothers, as I pointed out earlier. Female monkeys, and indeed females of most animal species, were seen as the equivalent of mothers, so that if you studied maternal behavior, you were studying all that there was to female behavior, as well as all there was to care of the young.

Females are of course mothers and behave as mothers, and female animals from many different species care for their young in many different ways. Yet the assumption that female equals mother is wrong on two counts, both of which limit our appreciation of what animals can show us. First, females do many other things besides act as caregivers to offspring. They may even behave in ways that are not characteristically feminine, which may be difficult for observers even to register if they are expecting nothing from females besides nurturing maternal behavior. For example, Marcy Lawton, a behavioral biologist from the University of Alabama–Huntsville, and two colleagues recently reexamined the work of ornithologists John Marzluff and Russ Balda. In a compassionate, critical, and ultimately hilarious send-up of the two scientists' study of sociality in pinyon jays *(Gymnorhinus cyanocephalus)*, a member of the crow family found in the southwestern United States, Lawton et al. note that although adult males in the species were said to fight very rarely and exhibit few tendencies toward social dominance, Balda and Marzluff persisted in a quest to determine the "alpha male" and to map a social hierarchy among the males in a group. Eventually they found what could be called a set of dominance interactions, though it was hardly dramatic and relied on signals such as sidelong glances among a small minority of the males. Lawton et al. point out that plenty of aggression occurred in the jays, and is documented in Marzluff and Balda's book on the biology of the species. It just didn't occur among males. Instead, females engaged in vigorous battles which even Marzluff and Balda acknowledged was "the most aggressive behavior observed during the year." But the chapter in their book on dominance relationships is dismissive of such female activity, saying only that "In late winter and early spring . . . birds become aggressive toward other flock members. Mated females seem especially testy. Their hormones surge as the breeding season approaches giving them the avian equivalent of PMS which we call PBS (pre-breeding syndrome)!"

Lawton and her colleagues wryly observe that "this is the kind of prose that made the publishers of *Ms.* Magazine rich" (p. 71), and go on to discuss how biases about what males and females are "supposed" to do can blind even good scientists—and they are clear about recognizing Marzluff and Balda as good scientists—to what animals might actually be doing. Females are not simply acting as mothers, or, with whatever counterpart of pre-menstrual bloating and depression can be imagined in birds, as the predecessors to mothers, waiting fretfully to lay eggs and get on with their mission in life. Do not assume, Lawton and company caution, that any interesting nonmaternal behavior must occur in males.

Don't get me wrong; I am perfectly aware that reproduction is indeed the ultimate goal from an evolutionary perspective, and that if the female jays (or individuals of any other species) fail to have offspring it does not matter if they spend the rest of the year composing symphonies or inventing computer chips. But female reproduction is about more than acting as a mother, and as we will see, winning in evolution does not mean always being the best caregiver, whether you are male or female.

The second reason that the assumption that females are equivalent to mothers is wrong is that fathers are important too, as are females other than the genetic mother of a juvenile. Long before New Age terminology started the trendy use of the word "parent" as a verb, males did a great deal of offspring care. In many different species of seahorses and their relatives the pipefish, the female places eggs in a pouch or other specialized structure on the male's abdomen, rendering him effectively pregnant. He fertilizes the eggs, and then may care for the developing embryos by aerating them, keeping harmful fungal infections away, and regulating the salt balance they experience. After the young develop in this protected site, they hatch, and the male gives birth to tiny fish that swim away on their own. Some species show monogamy, while in others males accept eggs from more than one female and the females compete among themselves for access to the nurturing male of their choice. While less common than female-only care of young, male parental care is seen in an enormous variety of animals, including many fishes, frogs and toads, insects, and a fair sprinkling of birds. In mammals there are even some conscientious fathers among primates; several different species of marmosets, diminutive monkeys that live high in the rainforest canopy of South America, have social groups consisting of a female and one or two adult males. When the female gives birth, usually to twins, the male almost immediately takes over every aspect of their care, including carrying them, one on each hip, every moment that they aren't being nursed by their mother. When it is feeding time, he hands an infant to the female and waits, apparently anxiously, until she is done, then hands her the other one. After the entire nursing episode is over, he regains both offspring with seeming relief, and all move on to a new place in the treetops.

Conventional wisdom holds that paternal care in mammals is not more commonly seen because females are the sex that produces milk, a fact that limits the degree to which males can take over rearing the offspring. Animals such as fish and frogs show no such constraints because either sex is equally capable of an activity like fanning eggs to keep them oxygenated, and hence male parental care is quite common in these groups. This begs the question of why male mammals have not evolved the ability to produce milk; they possess the basic apparatus, and in a couple of species, including a type of bat, they apparently do lactate. Enigmatic physiological details aside, however, the point is that there is no a priori reason to expect females to be better at care of the offspring. It just depends on the circumstances, and so equating females with mothers—or perhaps more aptly, equating mothers with females—ignores a great deal of the flexibility that is found in nature.

In situations where both parents participate in raising offspring, it is instructive to examine how much each parent contributes, and how this changes under different environmental conditions. Small songbirds have provided good test cases for ideas about whether mother or father does more, because after the eggs have hatched, in many species both male and female take part in the highly visible and easy-to-measure activity of bringing beakfuls of worms or other nutritious delicacies back to the nest. In fact, over the last decade or two a virtual cottage industry has arisen among ecologists, inventing ever more ingenious ways to experimentally manipulate the costs and benefits of bringing food to baby birds.

For example, males mated to more than one female usually do not bring as much food per chick as males mated monogamously, even after the number of nests is corrected for. On the other hand, in a few species, like zebra finches *(Taeniopygia guttata)*, males that are exceptionally attractive as mates (gauged by the color of their investigator-applied leg bands, interestingly enough) are able to loaf more than homelier individuals, who apparently compensate for their lack of physical charms by working harder to feed the offspring. Several researchers have swapped eggs among nests at the same stage of development, so that some nests have larger than average clutches while others have rather few eggs. In most species, the alterations go unnoticed, and the parent birds respond to the changed number of eggs as if nothing unusual had happened. When the eggs hatch, males and females may respond differently to the increased demand of an artificially enlarged brood; in a study of brood size manipulation of great tits *(Parus major)* in Switzerland, Heinz Richner and his colleagues found that males, but not females, increased the number of trips they made to and from the nest when more chicks were present. This had an unexpected side effect: the males also showed an increased incidence of infection with avian malaria, a mosquito-carried disease similar to the malaria seen in humans. It wasn't clear whether the males wore themselves out and hence became more immunologically vulnerable to disease, or whether they simply exposed themselves more to the vectors via the increased number of trips they made through the forest. Other studies of disease in birds, including barn swallows *(Hirundo rustica)* investigated by the indefatigable Anders Pape Møller at the Laboratoire d'Ecologie of the Université Pierre et Marie Curie in Paris, seem to suggest that the immune systems of the males actually change depending on their activities. Some diabolical scientists have even increased the effort required to fly to and from the nest by gluing weights onto the backs of birds; this too yielded different responses from males and females.

The general explanation of why the sexes respond differently, as we saw in Chapter 1, is that females usually maximize the likelihood of leaving genes in the next generation by ensuring the quality of the offspring that are produced. Females being by definition the sex that actually produces those offspring, they are clearly limited in this likelihood by the number of young they can successfully get to survive, and this number is relatively small for many species. Imagine the number of children a woman could produce, even at a hypothetical and unlikely-to-be-realized maximum rate: if she starts menstruating at, say, thirteen, and reaches menopause at forty-five, with a baby every year and a half to two years, that's an initially astronomical-sounding 16–22 children at a rough estimate. The number quickly stops being astronomical, however, if you compare it to the number of children that could be sired by a man operating at the same hypothetical maximum. Theoretically, it is difficult even to calculate this number, because it basically depends only on the number of women our imaginary Lothario could persuade to have sex with him at the appropriate time of their menstrual cycle, but it is clearly much larger than the maximum for females. In *Mother Nature* Sarah Hrdy contrasts a Brazilian woman named Madalena Carnauba, who gave birth to 32 children, with Moulay Ismail the Bloodthirsty, a Moroccan man who fathered 888 children in the early eighteenth century. Thus most evolutionary biologists conclude that males often benefit by simply seeking additional matings, but females gain little in the form of leaving genes in succeeding generations merely by increasing the number of mates. Instead they are more likely to have greater reproductive success by paying attention to the offspring in which they are already investing. These are rough generalizations, but for now they are a roundabout way of saying that females have more to gain if they focus on the well-being of the small number of offspring they already have.

The generalizations have another implication, one that is sometimes overlooked. The important point is maximizing reproductive success, or what evolutionary biologists call fitness—not simply having a lot of children, although these may sometimes, even often, be the same thing. It is those situations when the two are different where we find some of the most puzzling aspects of biology, and some of the points on which Charles Darwin said his theory of natural selection would stand or fall depending on how they were explained.

Why is parental care, either by males or by females, favored by natural selection? This seems like a facile question; after all, an individual's genes are contained in his or her offspring, and ensuring the survival of those genes is paramount from the standpoint of evolution. But there is an important distinction to be made here. Parental care is favored only insofar as it does help pass on genes. If some other behavior besides care of one's own young becomes advantageous under certain circumstances, mother-hood, or at least maternal care, goes the way of the dodo and the Edsel. And becoming a parent itself must be done judiciously or it too is worth-less, which is why we don't see many women at that ambitious maximum of twenty-some children. It is not maximal maternity, it is what you get out of it that counts. Rampant fecundity is worthless.

This means first of all that control of fertility is extremely important in nature. For females, reproduction is a continuum from ovum formation to egg production to laying/pregnancy to hatching/birth to whatever rear-ing occurs in a particular species. It is quite "natural" to stop this process at any time if it looks as if continuing with it will be detrimental to the passing on of genes. Even plants abort (yes, that is what botanists call it) developing fruits if environmental conditions are not favorable for their development. Mother mammals resorb embryos, mother birds desert nests. In the harsh inner deserts of Australia, when the rains have failed to come and vegetation grows scarce, the pouches of mother kangaroos of certain species begin to tighten at their openings. Inside the pouch is not only shelter but the mother's nipple and the source of milk. Eventually, if con-ditions continue to deteriorate, the young joeys can no longer enter their former haven, and may even be expelled from it. If they are still too young to survive without their mother's milk, they starve and die. No conscious decision, no heartbreaking dilemma, is required; mothers that did this and survived to reproduce at another time were more likely to pass their genes on than mothers that tried to keep their young alive as well as themselves, and lost the battle. And kangaroos are well suited for the unpredictable habitats in which they live: although the young joey is sacrificed, the mother has a tiny embryo, no bigger than your little finger, ready in sus-pended animation in her uterus. Triggered by the flush of verdant life that follows the rains, she can start her reproductive machinery at virtually a moment's notice and have another offspring in weeks, without waiting to

find a male and ovulate. If the going gets tough, however, the tough—
and those in it for the long haul—stop taking care of their children.

The idea, then, is not to take care of the helpless but to maximize one's
reproductive success, which can have some unexpected repercussions in
the evolution of behavior. Sometimes the repercussions mean that parents
stop taking care of their offspring, like the kangaroos and many other
species, or like the animals that kill their own offspring under stressful or
uncertain conditions. In laboratory animals this infanticide has been interpreted as abnormal psychotic behavior, an artifact of captivity, but at
least some of the time it is probably adaptive in nature, because rearing
young when life is harsh may be too big a gamble for it to be continued.
Taking care of children has to be selfish in evolutionary terms.

Another type of repercussion means not that parents stop caring for
their own young but that individuals actually start caring for someone
else's. One of the most stunning examples of this is the behavior that
puzzled Darwin when he was working on his masterpiece *The Origin of
Species:* that of the social insects. All ants and termites and many species
of wasps and bees live in complex societies, with distinct jobs for different
individuals and often physical differentiation as well. In the most familiar
example, the honeybee, a large queen produces hundreds or thousands of
eggs during her life. These are cared for by worker bees, all female, and
all remaining sterile while devoting their lives to feeding the queen's young,
fetching nectar and pollen from flowers in the neighborhood, repairing
and expanding the wax cells in the hive, and otherwise ensuring the smooth
functioning of the colony.

Darwin and many biologists before and after him found this behavior
unsettling because evolution should not favor giving up your own chance
at reproduction to help raise someone else's offspring. While being queen
is not all loafing around eating royal jelly bonbons, from the standpoint
of natural selection it is still the best job there is, because every one of the
maggotlike children being tended by the workers is carrying the queen's
genes. What do the workers get out of it? That is the part that had Darwin
concerned about the viability of his theory, based as it was on the differential survival and reproduction of individuals depending on their ability
to compete with others of their species. Sacrificing reproduction is a sacrifice like no other, and the social insects provide the best case of such
sacrifice in the animal kingdom. It is true that when honeybees sting a
perceived threat to the colony, they lose their lives, but suicide itself is

much less difficult to explain than lifelong sterility. People sometimes focus on the suicidal defense as the enigma, but a much more compelling one is the childless state of the defenders.

How the social insects evolved their complex systems of altruism is a still-unresolved topic. A partial answer appears to lie in the high degree of relatedness between the workers (they are often at least as close to each other as cousins, often sisters, and sometimes, through a peculiarity of wasp, bee, and ant genetics, even closer than vertebrate sisters), which means that helping rear the queen's offspring is helping some of the worker's own genes survive. Some of the rest of the answer may be that establishing a colony or reproducing on one's own is so risky and unlikely to succeed that even indirect reproduction via care of relatives is better than trying and failing completely, similar to the dilemma of the mother kangaroo.

Solutions like these also probably apply to the cooperatively breeding birds and mammals, species such as the acorn woodpecker *(Melanerpes formicivorus)* of the western United States or the dwarf mongoose *(Helogale parvula)* of Africa. In these animals, a breeding pair is helped by other individuals, sometimes young from previous years, sometimes not, in rearing the offspring that the breeders produce. The differentiation between the reproductive haves and have-nots is less clear than in the bees; no morphologically distinctive Queen of the Woodpeckers occurs. Indeed, a close look at the lives of these colonially breeding species suggests an undercurrent of strife beneath the apparent harmony, with female acorn woodpeckers sometimes throwing another's eggs out of the nest and subordinate dwarf mongooses occasionally having a litter of their own. Again, whether it pays to be maternal depends on the circumstances. If resources are scarce, a first-year bird is often better off staying with mom and dad, perhaps angling for a copulation or an egg of its own, than striking off and attempting to breed but failing and putting its own survival at risk. Furthermore, nonreproducing hangers-on may be allowed to remain on the territory in exchange for helping with child care, and they may gain valuable experience at taking care of offspring that will serve them well when conditions allow them to reproduce on their own.

How hard the individual tries for a proportion of the breeding in the group depends on the costs and benefits of doing so, in terms of the likelihood of leaving genes in future generations. If a subordinate female mongoose is in good condition and has some experience at taking care of infants, and if the environment looks reasonable, the odds are pushed in favor of her attempting to become independent of the group. Her older sister the breeder may then be more likely to "allow" her to have a small litter of her own if she stays, and will "permit" that litter to grow up with the help of the other group members. The help, after all, is worth something. Again, it is not necessary to invoke conscious decision-making; breeding sisters who were less aggressive toward healthy subordinates had higher reproductive success than more tyrannical ones who booted any reproductive sisters from the colony. From the standpoint of the subordinate, that raises the benefits of staying, along with the benefits of leaving. Everything has its price.

These complexities have led researchers to develop a body of theory based not simply on maximum fecundity and motherly devotion, but on the relative costs and benefits of different varieties of reproduction. This theory, called reproductive skew theory, attempts to quantify the genetic payoffs and penalties of various strategies: staying or leaving, having most of the young in a group all at once rather than a few at a time over several seasons, and so on. Fairly complex mathematics are required to calculate the outcome of such intricate social interactions, but the result may be what Steve Emlen, a well-known behavioral ecologist who spent many years studying the cooperatively breeding white-fronted bee-eaters *(Merops bullockoides),* a colorful African kingfisher relative, calls a "biology of the family." From my perspective, it means that we learn much more after we abandon a few stereotypes, like the assumption of single-minded sacrifice by a mother for her young.

The ultimate perversion of parental care occurs among the brood parasites, species which abandon their eggs or (less commonly) young to individuals from the same or different species. Cuckoos and the New World cowbirds are the best-known examples of this behavior, and some of the most incredible nature photographs I have ever seen are of a tiny reed warbler *(Acrocephalus scrirpaceus)* in England valiantly stuffing worms into the gaping mouth of an oafish cuckoo chick many times her size, the subject of study by Nick Davies and his colleagues at Cambridge University. Again, although this system has fascinated scientists and the public for centuries (our rather quaint term "cuckold" comes from the phenomenon in birds), ultimately the explanation from the viewpoint of the parasite, the cuckoo or cowbird, is that having someone else take care of your offspring yields higher reproductive success than taking care of them yourself. Why this is true for only a subset of species, and perhaps even more compelling, why and how the hosts put up with it, are the subjects of much ongoing research.

I will close with a more enigmatic example of offspring care, one that seems to cross boundaries of paternal care, maternal care, and brood parasitism itself. It occurs in an insect species so unnoticed that it lacks a common name, though it is sometimes called a golden egg bug; mainly it goes by its unwieldy Latin one of *Phyllomorpha laciniata*. It is a true bug, meaning that it has soda-straw mouthparts that suck the fluids from plant stems for food, and it lives in the Mediterranean area, where it has been studied by Arja Kaitala, a Finnish scientist from the far-northern University of Oulu. The bugs overwinter as adults and start breeding in the spring. When they mate, the females do something very peculiar: they glue their eggs onto the backs of other individuals, where they gleam a lovely golden yellow color. The puzzling part is that they do not simply glue the eggs onto the backs of the fathers, in a manner similar to seahorses. They do not—presumably they physically cannot—glue them onto themselves. A male is more likely to be found carrying eggs than a female, but sometimes those are the eggs that that male has fertilized and sometimes not. Females, too, carry the eggs of other females. All individuals carrying eggs are more likely than eggless bugs to be eaten by birds or other insects, presumably because the gold eggs render the caregivers much more conspicuous among the leaves. At the same time, eggs not glued to the back of another bug are almost certain to be either eaten by a predator or found by a parasitic wasp and killed that way. Even when eggs are on the back of a bug, the same parasitic wasps often locate them and lay their own eggs on them, using the bug eggs for nourishment of the wasp larvae that develop inside.

Life, then, is tough. Why take eggs unrelated to you from someone else? Why not find a better place to hide them? How did the eggs evolve to be so conspicuous, rather than cryptic? Why do both sexes carry eggs, and why do males harbor them more frequently? Is this parental care? Is it adoption? Is it parasitism? The system is part seahorse, part cowbird, and wholly a mystery. It illustrates the fuzzy boundaries that exist in animal care of offspring, and it suggests that we do well to escape the stereotypes and start trying to understand what animal family values really mean.

Four

DNA AND THE MEANING OF MARRIAGE

AT FIRST GLANCE ONE MIGHT not think that the need to prevent birds from eating farmers' crops has a great deal to do with the genetic basis of adultery or with male biases concerning animal behavior. It turns out, however, that a study undertaken over thirty years ago ended up paving the way for just such a connection.

Red-winged blackbirds *(Agelaius phoeniceus)* are familiar harbingers of spring in many parts of North America. The species gets its name from the red shoulder patches, snappily trimmed with yellow, that adorn the otherwise glossy black males. Females are drab in comparison, a somewhat mottled brown. In marshes and on lake shores throughout the United States and Canada, males settle into territories zealously defended against intrusion by other males and wait for females to arrive, which they do rather later in the spring. The females build nests among the reeds on the territories of males they prefer, and any particular male may have anywhere from zero to eight females in his area. Territories vary in their quality, and although a female does not feed on the territory in which she nests, the survival of her chicks depends on her choosing a good site, because many eggs and young are lost to predators such as snakes and raccoons.

Marshes are reasonably easy habitats to work in; the edges of the habitat are discrete and easy to map in a notebook, any birds displaying near the tops of reeds are conspicuous and visible, rather than hidden in the canopy of a rainforest tree, and the nests are accessible to the intrepid scientist by canoe, rowboat, or simply splashing around in the water wearing hip boots.

Once a nest is located, it is a simple matter to track the progress of its contents by checking it on a daily basis until the young have hatched, fledged, and left, or until the eggs or chicks disappear. All these conveniences have been readily exploited by ornithologists, who for the last many years have developed a niche discipline in the breeding biology of redwings, as they are familiarly called by birders. If males differ in the number of females they attract, and if the offspring in the nests where those females lay their eggs can be monitored, then it is a straightforward process to calculate the reproductive success of a male simply by adding up all the chicks that survive in his territory. Male fitness in the evolutionary sense thus mirrors the behavior that the scientist measures. This is not so easy to do with species where individuals are widely separated or where males, rather than visibly varying in the number of females they associate with, simply disappear if they do not find a mate.

One can also examine what the females get out of this arrangement. Red-winged blackbirds were instrumental in helping Gordon Orians, an eminent scientist now retired from the University of Washington in Seattle, to construct a theory about mating systems, the scientific term for the kinds of associations that occur between males and females in a given species.

Most commonly, mating systems are described as either monogamous, with one male and one female paired more or less exclusively during the mating season, or polygamous, with more than one member of one sex paired with a single member of the opposite sex. Polygamy in turn is divided into polyandry, where one female is associated with more than one male, and its converse, polygyny, with multiple females associated with a single male. Some people add a category of "promiscuity," implying that there is no observable structure to the associations between individuals; I am dubious about this designation, both because of its rather anthropomorphic loose-woman connotation and because I suspect that many cases that appear indiscriminate are instead merely not well understood. Strict monogamy is rare in the animal kingdom, probably because, as I noted earlier, males are expected to compete with other males for access to females more than the reverse, and because, as I discuss below, mating with multiple partners is often beneficial. A noteworthy exception appears to exist in many species of songbirds, not including the red-wings. In most other species of songbirds, though, a male and female share a nest and usually at least some of the work involved in incubating, feeding, and protecting the nestlings. Mammals tend to be polygamous, as do most other species, ranging from crabs to cockroaches, that have been studied in this regard. The mating system of a species has many important implications for other aspects of its life, including the population growth rate, the age at which individuals are expected to become mature, the degree to which males and females differ in size, and much more.

Orians studied red-wings himself, and was particularly interested in what females got out of sharing a territory with other females rather than distributing themselves so that each individual had one and only one mate. He suggested that being a second, third, or other additional female on a territory is worthwhile if that territory is so good that a share of the resources it provides, including the attentions of the territorial male, is at least as beneficial in terms of the number of offspring it allows a female to produce as a poorer territory. In other words, at some point it is better to join a female in a place that allows you to produce and protect three chicks than to go off on your own and settle somewhere that only allows you to have one or two. Orians termed this point the polygyny threshold, and the idea of evaluating costs and benefits of being singly mated vs. sharing with other females helped structure ideas about other types of mating systems in other species. The red-wings themselves continued to be scrutinized by biologists, who started performing experiments like reducing the showiness of males by blackening the red shoulder patches with ink to see whether females found them less attractive (apparently not, although they did seem to have a harder time keeping intruder males away). A great deal of research also went into exploring the hormonal basis of various behaviors in the blackbirds, like determining whether high testosterone levels were seen throughout the breeding season or just at certain key stages (the latter appears to be the case). Red-wings were—still are, really—a model system in their own right, at least for the study of mating systems in birds.

VASECTOMIES, GENES, AND GRAIN

While all of this was going on, however, there were rumbles of a complication on the horizon, although no one realized at the time what a profound change it presaged. In addition to their more admirable qualities, red-winged blackbirds are numerous enough in some places to be pests, because they eat grain from fields, sometimes in great quantities. Remember that the birds do not feed in the marshes where they nest, so they have to go somewhere else, and a flock of thousands of blackbirds can make a

sizable dent in a farmer's wheat crop. Scarecrows or their more technolog-
ically advanced counterparts aren't really effective, and shooting or poi-
soning large numbers of birds that many people think are symbols of spring
is bad PR, so in the early 1970s agricultural pest advisers were looking for
a means of controlling the birds without attracting a lot of unwanted
attention.

They hit on an idea that has long been promoted for human population
control: vasectomizing the males. Sterilized red-wings should still go
through all the appropriate behaviors like defending the territory; they just
wouldn't have any babies. The eggs would be laid as usual, because the
female couldn't tell the difference between a fertile and unfertile egg, but
they would never hatch. Perfect. Nondestructive, no consent forms
needed, just a skilled technician (and you thought those tubes inside a
human male were tiny) and some funding from the U.S. Fish and Wildlife
Service. So, in 1971, biologist Olin Bray and his co-workers cut the sperm-
carrying tubules of eight territorial red-wing males in Lakewood, Colorado. They timed the operation for the period just after nesting started,
in late May, and they removed the eggs from the females on the territories
of the vasectomized males so that the females would have to remate to
replace the lost clutch. It was this replacement clutch that was of interest;
the eggs laid immediately after the male had his vasectomy could have
been fertilized with sperm stored by the female before the operation. The
researchers then compared the fertility of these second sets of eggs to those
from nests in the territories of ten normal males.

Much to their surprise and dismay, the majority of the clutches produced
by the females from vasectomized males' territories were fertile. When the
study was repeated in 1972, just to make sure that the operation really
worked, the protocol was being observed, and there hadn't been some
peculiar mistake, the same thing happened: regardless of the status of the
territorial male's seminiferous tubules, females were laying fertile eggs,
which meant they were being inseminated by a bird other than the territorial male.

The introduction of the paper published on this research has a suspicious
mention of the possibility of "female promiscuity" compromising the results of sterilizing the males, though one wonders if this foreshadowing
only came about post hoc. The authors conclude, rather peevishly it seems
from reading between the lines, that this possibility was borne out in their
study. In the discussion of their results that appears at the end of the paper,
they ruefully cite several older papers that also noted the occurrence of

matings outside the pair bond; presumably these articles were discovered in hindsight. More isolated nests located farther from other male blackbird territories were less likely to contain fertile eggs, which supports the idea that the relationship between the females and male on a territory was not what it had appeared to be.

There are several interesting conclusions to draw from the study. From Bray's perspective, the point is that "female promiscuity" ruined the prospects of using male vasectomy as a method to control red-wing populations, though the paper speculates that in circumstances where territories are quite far apart, it might be possible to keep a rein on those loose females. I myself wonder why all the blame is laid at the feet of the females. It takes two, after all, to commit adultery, and yet the paper never refers to promiscuity on the part of males. This is an illustration of my objection to the term "promiscuity" in the first place; it suggests that the females are doing something wrong, something that should not happen, rather than allowing that our categorization of the mating system itself was flawed. But from an evolutionary point of view, the most provocative conclusion is that polygyny did not occur in the way we understand it.

After Bray's paper was published, he went on to examine other methods of controlling birds eating crops and, so far as I can determine, never thought about the morality of mating in blackbirds again. Other more behaviorally inclined biologists did note his findings, and the idea that paternity of eggs in a nest might be shared among several males was occasionally brought up, but for almost the next two decades the problem remained largely unstudied, mainly because of technological difficulties in determining who fathered which of the offspring in a group. Until recently, the best method available was called paternity exclusion. This allows one to rule out certain individuals as fathers, but it cannot conclusively determine that a given male actually *is* the father.

To illustrate paternity exclusion, let us consider the process as it used to be conducted in humans, using blood types. The molecules on the outer surface of blood cells are inherited from an individual's parents, and there are several different kinds. One of the most common classes of blood groups is ABO, with four blood types possible: Type A, Type B, Type AB, and Type O. Each person has two forms of the gene that produces these molecules, one from each parent, and the blood type is determined by the combination of these forms, or alleles. So a Type A person actually may be described as AA, because he or she has two alleles containing a code to produce the type A molecule. Some of the alleles are dominant, which means that even if you have only one representative of them, that is the blood type you exhibit. Types A and B both have this characteristic when they are combined with a Type O allele; if you are Type A, you may have in your genes an A allele and an O allele (AO), or two A alleles (AA). Both yield the same blood type. The same is true for the B allele. A combination of an A and a B allele results in a Type AB, but the only way to have Type O blood is to have both alleles be O, because if O is combined with either A or B the individual's blood type is that represented by the dominant allele.

Thus only two parents with at least one allele of Type O blood can produce a Type O child. Combined with the ability to test blood type simply in the laboratory, then, this information can tell you if it is possible that a given individual is the father or mother of another. If a baby is Type O, but the putative father is Type AB, it isn't possible for him to be the genetic parent, because only an AO, BO, or OO individual can donate the O allele that is necessary for a Type O child to inherit. We can exclude him as the father, hence the term "paternity exclusion" for the test. If, however, he had been Type A himself, we could not rule out the possibility that he carries an O allele and this was passed on to the child. If we knew his genotype was AA, that would also exclude him, but if he is AO, we cannot proceed any further. Obviously, however, many different men are AO, and therefore although we could not exclude an AO individual as the father, neither could we positively assign paternity to him. By using additional blood groups we could exclude greater proportions of individuals, but this technique does not allow us to simply identify a given man as the father of a given child.

A similar technique using other proteins found in the muscle tissue or other organs was applied to birds and various other nonhumans, and by the mid-1980s several studies had evidence of young in a nest being fathered by males other than the supposed mate of the female laying eggs in that nest. Studies of white-crowned sparrows *(Zonotrichia leucophrys),* indigo buntings *(Passerina cyanea),* mallards *(Anas platyrhynchos),* and acorn woodpeckers *(Melanerpes formicivorus)* all showed that a proportion of chicks ranging from 25 to 40 percent must have been the result of what we now call extra-pair copulations, or EPCs. But the methods were laborious and did not work for all species, and as with humans, the technique has the inherent limitation that one cannot sort through a group of males and unambiguously connect each of them to the various offspring they sired.

Beloved by prosecuting attorneys, viewed with suspicion by judges, and the source of a proliferation of acronyms and jargon (PCR, RFLP, micro-satellite, minisatellite, and Jeffreys probe, to name a few), DNA finger-printing, as it is familiarly called, has changed the face not only of criminal investigation but also of our understanding of relationships between the sexes. Unlike paternity exclusion tests, tests using samples of DNA from individuals of unknown relationship can determine, within a varying margin of error, whether a chick in a nest (or a baby in a hospital crib) was fathered by a particular male. The assignment of paternity does require making certain assumptions about such things as the gene mutation rate, and the degree of relatedness that exists within the general population, since using a highly inbred group of individuals who are already aunts, nephews, and cousins to many of the other individuals in the group will yield inaccurate results, but these need not concern us here.

The precision of DNA testing comes from the unique combination of genes we each possess. Because each allele is inherited from either the mother or the father, by examining an array of alleles from both the putative parent and other members of the population, one can match parts of the genetic makeup of an offspring with the individual from whom the offspring inherited those parts. It works better than blood groups because the number of individuals with the same combination of alleles, unlike the number of individuals with the same blood group, is extremely small. Using this matching with enough individuals and in a reasonably contained population, we can now assign paternity with well over 90 percent accuracy. There are still some limitations, of course; all the test can do is compare the chick (or baby) to individuals whose DNA has already been sampled. It isn't possible, at least not yet, to examine the DNA of a little boy and declare, ipso facto, "That's Elvis's son!" unless you already have access to Elvis's DNA. (An aside to the behavioral part of the story: One of the things this new tool and ones like it fostered was an interest in molecular biology on the part of ecologists and evolutionary biologists. All of a sudden we had a direct line on the link between proteins from DNA and evolutionary relationships, and the result has been an extraordinary flowering of work across disciplines. For example, a scientific journal with the title *Molecular Ecology* now exists; this would have been undreamed of twenty or thirty years ago.)

But back to the red-winged blackbirds. A Canadian scientist named Lisle

Gibbs, who had been working on the ecology of Darwin's finches in the Galapagos Islands, began to take an interest in the new field of molecular evolution after he received his Ph.D. He decided to use DNA fingerprinting techniques on a handy species: the red-wings nesting in a marsh not far from Queen's University in Ontario. He and his co-workers mapped the territories of the males, noted the number of females that settled on each territory, and then counted the eggs and chicks that each nest produced. They then took tiny samples of blood from all the individuals on the site and typed their DNA. Birds, incidentally, make better subjects for this analysis than mammals if one is limited to obtaining only a small amount of blood, because their red blood cells, unlike ours and those of other mammals (with the peculiar and completely inexplicable exception of camels) are nucleated. A human red blood cell looks like a doughnut, because the nucleus is absent after the first stages of cellular development. Because the nucleus is the part of a cell that contains the chromosomes and hence most of the DNA, bird blood is easier to use for typing than mammalian blood. When it is complete, the analysis yields dark-colored bands of dyed protein strung across a gel, or rubbery matrix that allows the proteins to pass through it. The different proteins are associated with different alleles, and because a chick can get alleles only from his or her parents, one can trace the parental source of each band in each chick's gel, given the assumptions I outlined above.

When Gibbs lined up the gels, he discovered something startling: many, if not most, of the chicks in a given blackbird territory were fathered by a male other than the territorial one. The paper he and his co-workers published shows the map of the territories with two numbers for each nest: the number of chicks fledged from that nest, and the number sired by the territorial male. In the old days before DNA testing, these two numbers were assumed to be the same, Bray's troubling results notwithstanding. But Gibbs's results were too striking to ignore; the two numbers, far from being identical, were not even correlated. Males with four chicks on their territories might have fathered five others in the marsh, while males with ten chicks on their territories might actually have fathered only one. It wasn't just a matter of a couple of chicks here and there, it was as if the entire method for calculating reproductive success, that cornerstone of evolution, was shown to have a foundation of sand.

Telling my animal behavior students about this research triggers the strongest reaction they have to anything I teach them in the course. It is as if I were telling a group of astronomers that the earth is, after all, the center of the universe, orbited by the sun. They are horrified, and seem to respond as if to news of a death, or terminal illness, with those characteristic stages described by the psychologists. Denial: surely this was an exceptional case, or a peculiar year, too hot, too dry, too cold. Maybe they hadn't sampled enough males, or enough marshes, or enough chicks. I point out that none of these appeared to be true, and that furthermore, extra-pair mating and offspring resulting from it now appear to be widespread in birds. Then comes anger: the students are actually angry, some of them, though they are uncertain of the object of their emotion. When they do express it, half-joking, however, it seems to be directed at the female blackbirds, a bit like Olin Bray's denouncing the females as promiscuous while indulging the males in their sexual excesses. None of the students seem to go through the bargaining stage ("If the blackbirds behave themselves I will never whine to my teacher about another exam grade"), which I sometimes regret, but eventually they, like the scientific establishment, make it to the final stage, acceptance.

Incidentally, the data-gathering processes here also throw into relief one of the most pervasive myths I deal with in teaching, that one handles baby birds and then returns them to the nest at great peril, because the mother will "smell a human" on her offspring and reject it. Most birds have a relatively poorly developed sense of smell, and biologists have in fact capitalized on this obliviousness of avian parents to make wide use of cross-fostering experiments, in which chicks are swapped among nests to determine the contributions of parental genes versus parental behavior to their development. In most cases, parents continue to feed an alien nestling as if it were their own. Students have always heard the tale, however, regardless of where in the world they grew up, and are amazed to have it refuted; when asked what was most memorable about a vertebrate biology class I had taught, virtually all of them named the debunking of this myth. Some suggested that their mothers had made it up it in order to avoid children bringing home needy and messy new pets, but this cannot explain how all of them heard the same explanation, or why even those mothers tolerant of frogs, caterpillars, and dogs would always draw the line at baby birds. I have often wondered what a developmental psychologist studying college age students, presumably well past the age of independence, would make of this fascination with the possibility of maternal rejection.

To return to extra-pair copulations, however, I should emphasize here that Gibbs's paper was not the earliest example of its kind, merely one I know about because he is a friend of mine from graduate school. Other studies can be joined to it, and they have provided, within an astonishingly short time, a picture of animal relationships radically different from the one we used to have. Hundreds, maybe even thousands, of papers published in scientific journals over the last fifteen years document extra-pair paternity in birds from every continent, every habitat. In fact, it is no longer enough to replicate a study of extra-pair paternity in yet another species; students starting their Ph.D. degrees would be advised against such an undertaking unless they could contribute a new twist on the subject. A recent paper pointed out that EPCs are now known to occur in every avian family. That means ducks, warblers, woodpeckers, wrens, orioles, the lot. This is that same group held up as a model of monogamy just a few short years ago. It was a real revolution, and it took place within less than a decade.

Two questions now emerge. First, why do people find this so shocking, not just a surprise, but a disturbing revelation about the animal world and potentially our own relationships? And second, what does this finding mean for the science of animal behavior and our notions about mating systems?

IT ISN'T CHEATING IF THERE ARE NO RULES TO BREAK

The answer to the first question is fairly obvious; people are shocked by the blackbirds because we identify with animals, particularly attractive or cuddly animals, as I discussed earlier. Terms including "adultery," "infidelity," "betrayal," "cheating," "fooling around," and more have been applied to findings like those of Lisle Gibbs in the popular press, and sometimes the scientific literature is not far behind. Birds have always looked so admirable in their industry during the breeding season, the way the male and female rush back and forth to their demanding brood of chicks. And there is something uncannily amusing about mere birds being placed in the same situation as millions of modern-day husbands, eyeing a child warily, making uneasy jokes about the milkman. It is tempting to go on to conclude that if warblers, robins, and other models of monogamy are doing it, we have got to admit that extra-pair copulation, or adultery—or whatever term you prefer—is natural and expected, and maybe we should stop making such a fuss about it and resign ourselves to our evolutionary heritage.

This conclusion, however, is precisely what I am arguing against in this book. The birds aren't "cheating," they are just doing what they do, and they did not invent the rules about the pair bond between a male and female, we did. It isn't cheating if there are no rules to break. If we try and use their behavior as a model or justification for our own, we not only run the risk of making decisions about our morals on very shaky grounds, we miss what is interesting and vital about the animals' own behavior.

This is not to say that we cannot see a great deal about the influence of social expectations on scientists by examining their work on animal behavior. For instance, some of the literature on EPC is interestingly divided as to whom it portrays as the active party. Initially, there seemed to be two approaches, neither one particularly favorable to females. Either males were roaming around and taking advantage of hapless females waiting innocently in their own territories for the breadwinner males to come home with the worms, or else females were brazen hussies, seducing blameless males who otherwise would not have strayed from the path of moral righteousness. Bray's "female promiscuity" label is just one example. A paper published in the prestigious journal *Nature* refers to young in warblers as "illegitimate," as if their parents had tiny avian marriage licenses and chirped their vows. That some scientists in our society take this view should come as no surprise to us; after all, it was Hester who wore that scarlet letter, not her partner, and the double standard of judging adultery in humans has received much attention from sociologists and feminist scholars. I worry, however, that we allow our prejudices to influence our interpretation of what we see the animals doing, and analysis of extra-pair copulation is a prime example. Another mating system–related case illustrating the influence of bias comes from a 1972 paper on the Tasmanian native-hen *(Gallinula mortierii)*, a bird with a rather complex set of relationships between the sexes, including what appears to be polyandry, multiple males associated with a single female. The paper refers to this behavior as "wife-sharing," but I have never seen polygyny, its mirror image, called "husband-sharing." Making the males the active parties (they "share" the female, as if she were a six-pack of beer) may reduce the likelihood of noticing what the females do, of seeing things from their point of view.

In response to the second question, the most productive approach to this revolution in mating systems is rather to explore what extra-pair copulations mean for both males and females and what environmental factors might favor or disfavor them. This avenue has indeed been taken by many researchers, particularly recently; I do not mean to suggest that all scientists are unable to take a balanced point of view. Several papers from the 1990s evaluate costs and benefits of EPCs to both sexes. In one of these, published in 1998, Marion Petrie and Bert Kempenaers remark, "In most bird species it is likely that females control the success of a copulation attempt and of transfer of sperm," a statement that would have been unlikely to appear in print a couple of decades ago. Other researchers are trying to determine whether the population density of individuals within a species, the tendency for a species to migrate rather than to breed and winter in the same place, or some other aspect of a species' ecology may influence the opportunity for extra-pair mating.

It is likely that males and females derive different benefits from such mating, because of the different sorts of limitations on male and female reproductive success. Settling on a territory with one male and mating with another can allow a female to increase the genetic diversity of her offspring. In an unpredictable environment, having chicks with as many different characteristics as possible increases the likelihood that at least some will survive and reproduce themselves, a rather literal version of not putting all your eggs in one basket. This may be particularly true if circumstances do not allow many females to be associated with males possessing especially favorable characteristics; at least some of the offspring a female produces can be sired by the best male around, assuming there is some way of distinguishing this hero from the rest. Variations on this "good genes" idea abound, and another study by Petrie and two co-workers found that in species where more genetic variation exists, extra-pair paternity was more common, an intriguing and heretofore unexpected connection. Mating with more than one male can also ensure against infertility or low sperm counts in a given mate, or make a female's chance of re-mating higher if she loses her first mate because she has already established a bond with another male.

Engaging in matings with males other than the one whose territory she shares may also have costs for the female. Although little-studied in birds and other nonhumans, sexually transmitted diseases are a risk that increases with every sexual partner, and bird sex may be particularly nasty in this regard. Unlike mammals, most birds lack a penis or other intromittent organ, and the genito-urinary and digestive systems all empty into a common chamber, called a cloaca (this structure is also present in reptiles and amphibians, as well as some fish). Fertilization is accomplished by contact between the birds' cloacal openings, with sperm being released by the male into the cloaca of the female in a hurried operation that, at least to a human observer, doesn't look like much fun. (The process was referred to as a "cloacal kiss" by some overly poetic naturalist many years ago, but despite watching numerous birds have sex, I have never quite seen the analogy.) In any case, this set-up means that in addition to the usual exposure to the bacteria, viruses, and other germs associated with the genital tract, sex involves contact with the digestive and urinary systems, and we know what hotbeds of microorganisms those can be; this is why you wash your hands after using the toilet. I have always wondered if animals with a cloaca run a greater risk of sexually transmitted diseases than those without, but to my knowledge this question has never been explored. Cloaca or not, however, the more sexual partners you have, the higher the risk of catching a disease from one of them.

This cost also applies to males, of course, but females have other risks as well. The major one is considered to be the loss of a male's help in offspring care if, from his perspective, there is some chance that the young have been sired by another male. This risk appears to vary among species and in different circumstances, but may have shaped part of the inherent conflict that occurs between the sexes. In addition, looking for some action on the side takes time and energy, and many birds are already working flat out during the breeding season, finding food, chasing away predators, and keeping the chicks warm. Taking time from any of these activities may mean less well-fed offspring that might not successfully compete with other birds when they grow up, a potentially large risk indeed.

In just the last few years, studies examining these costs and benefits for the sexes have proliferated. The proportion of offspring sired by extra-pair males appears to vary from 0 (real, true, till-death-do-us-part monogamy is still seen among snow geese and some sea birds) to a whopping 90 percent, the latter in a species of fairy wren (a group of birds in the genus *Maluru*), brilliantly colored tiny birds from Australia. Do species with showy ornaments like these have a greater tendency toward EPC? What about variation in paternal care? Can males adjust the amount of energy they expend on offspring depending on the likelihood that they share genes with all of the young in the nest? What characteristics are preferred in extra-pair partners, and do those differ from the traits preferred in more long-term mates?

Answering questions like these will keep scientists busy for many years to come. We now refer, somewhat awkwardly, to a species' social mating system (for example, males and females appear to have a single partner, hence social monogamy) as distinct from its genetic mating system (in fact, many young are the product of extra-pair mating, hence genetic polygamy), recognizing that these are often not the same in a single population of animals. It is important to remember that these are our categories, not theirs, and we cannot expect the animals to fit neatly into them.

The news about extra-pair copulation shook the world of behavioral ecology, and with luck it will shake some stereotypes down with it, in particular some male-centered stereotypes. In a recent paper, feminist scientist Patty Gowaty pointed out several biased practices in the study of animal behavior that a less male-centered viewpoint helps to shatter, such as examining female behavior only with respect to males, or seeing females only as either virgins or whores, a point I echoed above. With characteristic bluntness, she also suggested that some ideas about female behavior in animals "would seem to have more to do with the nature of some men's minds than most females' lives."

In an even larger sense, however, I think we are groping toward a completely new view of animal interactions. Two ideas have been longstanding in the literature about animal mating, particularly if invertebrates are included: females often mate with more than one male, and polyandry is rare. Those two notions seem diametrically opposed, and yet they both seemed to be supported in nature. Multiple mating by females is an undeniable empirical observation, but it has been difficult to explain, in part because polyandry, the long-term association of a female with more than one male, both seemed theoretically unlikely to evolve and has been observed in only a handful of species. Polyandry looked implausible because at first glance it seemed detrimental to males and nearly pointless for females to maintain more than one mate; two males cannot get a female any more pregnant than one, whereas multiple females can each be inseminated by a single male. Hence systems like that we supposed the redwings had seemed understandable, whereas those found in birds like jacanas (genus *Jacana*), tropical water birds also called lily-trotters from their long slender toes useful for stepping on lily pads, were less common. Female jacanas are larger and more brightly colored than their male counterparts, and they compete for groups of males that are herded onto the female's territory. Each male mates with the female and cares for the eggs that she leaves with him, almost like the seahorses described in the previous chapter. A few other shorebirds show this kind of role reversal. Why these and not other species are polyandrous has remained unclear, but the rarity of polyandry seemed undisputed.

And yet—females of a lot of different species, including jacanas and now it seems red-winged blackbirds and many others, mate many times, with the same or different males. Often those matings are solicited by the female herself, in animals ranging from butterflies to rhesus monkeys. If females are coy and passive, if polyandry has no benefits, how do we explain this? I think the answer is that strict polyandry in the form of a social mating system may be unusual, but that we have allowed our categories to blind us to what the animals are doing. From the animals' perspective, or more correctly from the perspective of evolution, it does not matter what the system looks like. The only thing that matters is leaving genes, and if mating with several males helps, a female will try to do so. I do not know if we need new terminology, or just a recognition that the old terms need some stretching to fit reality. In either case, the next interesting part of sexual relationships comes about because what benefits a female is not necessarily what benefits a male, and vice versa.

Five

THE CARE AND MANAGEMENT OF SPERM

EVEN THE MOST HARD-CORE depictions of human sexual activity stop being graphic after ejaculation and/or orgasm. The rest of the behavior—having cigarettes, rolling over and going to sleep, the afterglow—simply does not receive the same kind of prurient attention as the events leading up to the release of sperm. The little guys are in there, it will be ages before we know if she is pregnant, and there is no point in going on about sticky spots on the sheets or the withered appearance of a flaccid organ. With animals, too, scientists used to assume that copulation was the end of the story, and cute Woody Allen-esque depictions of nervous sperm cells to the contrary, all that needed to happen after the male ejaculated was for the swimmers to do their thing. Swimming, after all, is not really a team sport, so each individual cell just tried as hard as it could to be The Winner. Only a male could win, according to this view, so males were therefore the only interesting players.

This chapter is about why that assumption is far from true, and about how the action inside a female after copulation is as important from the standpoint of evolution as what goes on during even the most energetic coupling. Furthermore, it is about how, once again, a female perspective suggests some reinterpretations of what we thought of as the rules. Females do more than simply provide a pool with the medal at one end of the swim lanes.

As we have seen, a female often mates with more than one male, sometimes in rapid succession, which leads to the possibility of sperm from more than one male simultaneously occupying her reproductive tract. The opportunity that this affords for one of the males to succeed in fertilizing the female's eggs while the others' sperm fails has not escaped biologists, particularly those working on insects, among which males often have unusually elaborate genitalia and females unusually elaborate reproductive tracts. When these structures were examined under the microscope, they revealed the potential for complex movement of sperm before fertilization. Geoff Parker, a British biologist at the University of Liverpool, wrote a fascinating paper (often referred to as "seminal" by those unable to resist the pun) in 1970 in which he outlined mechanisms for what he called "sperm competition." Parker pointed out that males can certainly continue to compete for fertilization after copulation has occurred, they just do it via the one-celled messengers of themselves they leave as a consequence of mating. The process would, he recognized, lead to different kinds of selection on males. On the one hand, attributes that allowed a male's sperm to win at fertilization by circumventing the efforts of other males' sperm would be favored by selection; on the other, males that could prevent a female from remating in the first place would do well because they would avoid the whole problem from the start. Parker and his colleagues developed a series of theoretical descriptions with fairly advanced mathematics of what would favor different sperm characteristics depending on a variety of circumstances.

The process can be studied in any organism, and indeed probably occurs in many if not most of them, including plants, as long as the opportunity for multiple mating exists. It has been most closely examined in birds and especially insects. A few terms are worth explaining. First, "sperm precedence" refers to the order in which sperm fertilize the eggs relative to their entry into the female; first-male precedence thus follows a "first-in, first-out" rule, while second-male precedence follows the converse, "first-in, last-out." "Sperm mixing" may also occur, when the sperm from different males sloshes around together inside the female tract. The proportion of eggs fertilized by the second male is called the "P2 value," and can vary from 0 (the first male fertilizes them all) to 1 (the second male fertilizes them all), with a lot of variance in between. A P2 value of 0.50 would mean that neither male's sperm had precedence, and either one was equally likely to father the offspring.

Most early investigators studied sperm competition by looking at its products, the offspring, because obviously direct examination of sperm-sperm interactions is technically extremely difficult. They used a technique in which two males were mated to a female in a known order. One of the males, either the first or the second, was given X-ray irradiation, which has the same effect on the reproductive organs that radiation of humans produces: the male becomes sterile, though he is still capable of behaving normally in mating. Incidentally, my friends who have done this sort of work say that you just don't know what an awkward question is until you are sitting in the radiology department of a hospital with a box of beetles on your lap and someone asks you why you are there.

Once a group of males is thus rendered sterile, they are used in the aforementioned experiment. Let us say that twenty females were mated to two different males each, one sterile and one normal and fertile. Forty different males would be used to prevent the results being unduly influenced by some odd characteristic of a particular male. The offspring from the females would then be counted. Assuming the female cannot internally distinguish between killed and viable sperm, the number of young she produces will tell us what pattern of sperm precedence occurs in this species, because she will use sperm from either the first or second male or both to fertilize the eggs. If the first male she mated with was sterile, but she produces a normal number of offspring, then second-male sperm precedence must be operating and the P_2 value is 1. If the second male was sterile, and she produces the same normal number, then first-male sperm precedence occurs, and the P_2 value is 0. If she produces half the normal number of offspring, regardless of which male was sterile and which fertile, sperm mixing occurs and the P_2 value is 0.5. These days, with the advent of DNA testing to determine paternity, irradiated males are not as essential, but P_2 is calculated in a similar way.

Now, sperm precedence is not synonymous with sperm competition. Sperm competition occurs whenever the ejaculates from two or more males compete, which may happen through a first-in, last-out mechanism or otherwise. Looking at the various mechanisms by which sperm from different males may find themselves inhabiting the same reproductive tract is instructive in showing how a male or female bias can change interpretations.

First, the frequency and timing of copulation affect the likelihood of sperm competition. If a female mates with more than one male, the timing of those matings influences which male, her "own" or an extra-pair, actually fertilizes the eggs. Female chaffinches, small European songbirds *(Fringilla coelebs),* produce a distinctive vocalization that attracts males, and they solicit 93 percent of the copulations that Oxford scientist Ben Sheldon and his co-workers observed. The chaffinch is a socially monogamous species, but extra-pair copulation is fairly common, and females are more likely to solicit extra-pair males when they are close to their fertile periods, making it more likely that these matings will result in offspring. Timing is everything. In addition, if mating with the same individual occurs quite frequently, the probability of that individual swamping the effect of another's ejaculate increases, and thus frequent sex beyond what is strictly necessary to produce a fertilized embryo is common among animals. Contrary to popular belief, animals do not as a rule mate in a quick, businesslike fashion and get it over with as fast as possible until the next fertile period. Mammals such as lions as well as many seabirds and birds of prey mate tens or even hundreds of times for each fertilization. The true champions of prolonged sex are the insects, however, with several species, including the relatives of bedbugs, staying coupled for hours or even days. Frequent mating often occurs at the behest of the female, one point that in my opinion casts doubt on the idea of female passivity during sperm competition.

Interestingly, frequent copulation is found in conjunction with two other characteristics, one behavioral and one physical. The behavioral characteristic is called "mate guarding," and describes a common occurrence among animals, both vertebrates and invertebrates, in which a male either remains with the female after copulation or follows her closely, interfering with any subsequent attempts to mate with other males. If it is successful, the behavior prevents sperm competition from beginning at all. As you might expect, however, mate guarding is more common in species where polyandry also occurs, which suggests that the guarding is not altogether successful.

The physical correlate of frequent mating is ejaculate size and testis size, both relative to the size of the male, and this is the second mechanism of sperm competition I wish to mention. Most people are familiar with the slang phrase "hung like a horse," but the intended implications for virility are realized in many animals that lack what most of us would call a penis, including many species of butterflies. In fact, the evolutionary function—and value—of a large penis is unclear; in mammals a longer penis doesn't get sperm closer or faster to the fallopian tube of the female, where fertilization occurs, and in insects the penis usually does not reach to the sperm-storage sites deep in the reproductive tract. Testes and ejaculates are another matter, however, with larger testes generally producing larger ejaculates. In the Pieridae, a family of butterflies that includes the familiar cabbage white butterfly, ejaculates are significantly larger than in a family with similar body size but a less polyandrous mating system, the ironically named (at least in this context) Satyridae. More detailed studies on animals ranging from other butterfly groups to primates show the same relationship: the more males a female is likely to mate with, the larger the ejaculates produced by those males. You do need balls to play, at least if you want to play with the other males. More sperm means a greater ability either to fill the female's reproductive tract and prevent later sperm from entering, or to literally swamp out the effect of other males' ejaculates if they got there beforehand.

The third mechanism by which sperm can compete is by physically elbowing, so to speak, the other sperm out of the way, and to do this they need help from their conveyor, the male reproductive organ. Jonathan Waage of Brown University was one of the first scientists to study this in detail, using a species of damselfly common in many parts of eastern North America. He discovered a shovel-like structure on the end of the male damselfly's penis (more accurately called an intromittent organ or, even more pedantically, an aedeagus). Through a series of elegant experiments, Waage determined that the structure is used to displace the sperm left by previous mates of the same female, after which the male's own sperm is deposited in its place. Photographs of male genitalia from this and other damselfly and dragonfly species taken at extremely high magnifications show a bewildering array of spines, knobs, and scoops, at least some of which apparently function to remove other contenders from the fertilization ring. In some species, a male tamps down the sperm from previous matings, rendering it less accessible, before overlaying it with his own. Mating plugs, sticky corklike objects composed of hardened mucous or other secretions, are left in the female reproductive tracts of many mammals and some invertebrates after mating; they were thought to serve as chastity belts by preventing entry of successive male organs, but their function is now debated.

Finally, sperm may compete by altering their shape or size or by making themselves into a team, contrary to the solitary swimmer image I invoked above. In his original paper, Parker noted that sperm within an ejaculate must compete not only with sperm from rival males, but with each other, and therefore any attribute making one individual cell better able to succeed should be subject to selection. One might expect, then, both variation that helps each cell do better and variation that helps the male producing the entire ejaculate to succeed in fertilization. Indeed, sperm morphology is amazingly varied, and among the insects alone many different types occur, including some with multiple flagellae, the whiplike organs used to propel the sperm through the medium. Among virtually all butterfly species, two types of sperm are produced: eupyrene, which has a DNA-carrying nucleus and is capable of fertilization, and apyrene, which is smaller and has no genetic material. In the humble fruit fly *Drosophila,* what are referred to as giant sperm are common in certain species, with some ranging up to twenty times the length of the male producing them. Alert readers may wonder how such large cells fit inside either sex. The answer is that the sperm "tails" are extremely thin and coiled, so that, much like your intestines, they can fold up into a relatively compact volume. The flies are tiny, of course. To put this into perspective, to achieve a similar feat, a human male six feet tall would have to produce sperm cells that could span a sizable portion of a football field (to carry on with the sports analogies that inevitably seem to accompany discussions of sperm competition). Humans do show some variability in the size and shape of sperm cells, albeit not as spectacular as that of the *Drosophila.*

What is the adaptive significance, if any, of all this variation? Some scientists have suggested that the different sperm morphs have different functions, with only a tiny minority of sperm actually able to reach the egg. The other sperm cells act as blockers of rivals or helpers of the real champs (for example, they make it easier for the fertilizing sperm to move through the female reproductive tract), but are themselves sacrificing their own chances for survival. This idea has been applied to humans, and the nonfertilizing sperm dubbed kamikaze sperm, for obvious reasons.

It is a colorful theory, but the evidence, at least as it pertains to humans, is weak at best. Many sperm that appear nonfunctional didn't get that way through a plan; they represent errors in manufacturing. And the evidence about what is retained vs. rejected in women's reproductive tracts comes from studies of what is exuded after sex in a group of volunteers who may or may not represent the general population. In butterflies, the theory that seems to have the most support is that the non-nucleated sperm cells are cheaper to produce and hence may act as "filler," allowing the male to counter other ejaculates with quantity if not quality. Short and long sperm of other insects may vary in mobility, with some more useful in the short term and some having higher survival ability after sperm are stored for some time.

CHERCHEZ LA FEMME

So what is wrong with this picture? The study of sperm competition seemed to be getting along swimmingly, if you will permit me a pun of my own. What was wrong with it began to be noted by researchers in the late 1980s and early 1990s, and what they noticed was that everyone was acting as if all the intrigue and sword-crossing was taking place in a Tupperware container with an ovum at one end. The female in whom the sperm competition was occurring did not seem to come into the picture except as a convenient vessel, the arena for all the action. Sperm competition seemed to be exactly analogous to the physical combat of the male organisms themselves, except that the sperm cells lacked the analogue of fists and had to make do with other sorts of assault. The dogma ignored the possibility that the ejaculates of more than one male got into a female in the first place via some action on her part, or that the elaborate reproductive tracts of many species were the result of selection on the females for differentiating among sperm. Some behavioral biologists, particularly primatologists, had been noting the existence of multiple mating by female monkeys throughout the 1970s and later, but their observations did not make it into literature about other animal groups. The differential fertilization of offspring by various males had only to do with male ability to get to the egg, by moving faster or by kicking the other sperm out of the way.

In short, scientists had not usually looked at sperm competition from a female perspective. It was a situation reminiscent of the days of the Greek philosophers, when men were thought to plant a seed, the homunculus, which contained all the information needed to make a human being, in the fertile but vacuous ground of the female. All a woman did was supply the soil, the growing medium, for this seed; the kind of fruit it bore was entirely up to the male, and the homunculus got its name because for several hundred years scientists claimed they could see a tiny person coiled up inside the sperm cell. (I have always wondered whether they saw tiny flies inside fly sperm, tiny porcupines in porcupine sperm, and so forth.) Women thus merely harbored the fetus; they had no part in creating it. Feminist critics have more recently pointed out the blatant sexism of this view, aside from its obvious wild inaccuracy, and I will not belabor it here. I think, however, that a milder but similar scenario applies to the formulation of sperm competition theory. Females were imagined to be minding their own business while a cellular version of the video game Mortal Combat went on inside their reproductive tracts, and they were assumed to have no influence over either the steps leading up to the conflict or its outcome.

In their introduction to a scholarly book chapter on sperm competition in insects, Leigh Simmons and Mike Siva-Jothy declare, "Insects are predisposed to high levels of sperm competition because females show a propensity for multiple mating, and maintain sperm in specially adapted sperm-storage organs (usually termed spermathecae)" (p. 342). This is not a controversial statement or one that calls attention to a little-known fact. Any researcher into sperm competition for the past quarter-century would agree with it. Yet both of these attributes making sperm competition likely are characteristics not of males, but of females: multiple mating and organs that store sperm. Why, then, would it not be logical to assume that selection had acted on the females, and males were at the very least competing with other males in an arena that was not of their own making? If males can remove sperm of other males, why can't females? Whose organs are we talking about, anyway?

SETTING THE RULES OF THE GAME, OR FEMALE ARCHITECTS

Let us return to some of the mechanisms for sperm competition I discussed above, and see if a female perspective—not necessarily one by females, just one that takes female costs and benefits into consideration—changes some interpretations. First, the frequency and timing of copulation, which are often, if not always, under female control. In the chaffinch, mentioned above, females initiate sexual activity, and they clearly influence which male fertilizes their eggs simply by soliciting copulation at particular points during the reproductive cycle. Many species of animals have similar patterns, in which females are active participants in mating, sometimes initiating, sometimes refusing. Completion of the sex act generally requires cooperation from both partners; in crickets, for example, the female must mount the male from behind and position herself to receive a sperm package, or spermatophore, the end of which is threaded painstakingly into her genital opening. The sperm then flow into the reproductive tract over a period of perhaps half an hour or more, depending on the kind of cricket. Females can simply walk away at any point during the process, and there is little that males can do to stop them.

The crickets also show mate guarding behavior; after the spermatophore is transferred, a male remains close to the female, touching her with his antennae and attempting to prevent her from removing the spermatophore, which she sometimes tries to do. The idea is that he is ensuring that his sperm fertilize her eggs and that no other males try to mate with her. When I first heard about this as a graduate student studying cricket behavior in the mid-1980s, it struck me as odd that no one thought about the female's role in this, about whether it was perhaps a challenge on her part. If she succeeds in removing a spermatophore before it has finished draining all of its contents, the male must produce another one if he is to fertilize the maximum number of eggs, a demanding task not all males are capable of (males infected with a protozoan parasite, for example, cannot replace spermatophores very readily at all). Females therefore may gain by remaining around the male and goading him, in effect, into performing a behavior that only a good-quality male can execute successfully. Calling it mate guarding makes it sound as if the female just stands there making like the crown jewels. In other animals females seek other males as mates during mate guarding. Why don't we call it "mate challenge" behavior, or "mate seeking" behavior?

The idea that there is often little that males can do if females do not cooperate in a mating attempt is also germane to sperm competition. Females influence the fate of sperm if they do not mate with a male at all, which is fairly obvious. But even if a male forces a copulation they can still have a pivotal role in the use of sperm they receive. A recent and very intriguing example was discovered by Nancy Burley, a biologist who since the 1970s has been studying zebra finches *(Taeniopygia guttata),* those small cage birds with the monotonous beeping vocalizations and the red bills. Burley noticed that male zebra finches often forced copulation on females with which they were not paired; in fact, 80 percent of all extra-pair copulations were aggressive, which she defined in a rigorous and repeatable way. These matings never resulted in any offspring, which is interesting by itself. Even more interesting, though, was that 28 percent of chicks in the aviaries were from the remaining 20 percent of EPCs that were not forced, an astounding success rate. What were the females doing to influence the fate of sperm from different males? No one knows. The cloaca of female birds is clearly capable of some sophisticated maneuvering; in several species, females have been observed ejecting sperm after a copulation. The organ's structure and function has, however, been relatively little studied by scientists. Similarly, sperm storage organs in many female insects at least potentially would allow females to use some sperm and leave other sperm alone.

I must hasten to add here that these findings do not suggest that females, including human females, can "decide," consciously or not, to become pregnant. The zebra finches, and perhaps other species, simply illustrate what should have been common sense: if selection acts on males to enable them to overcome competition and resistance and fertilize eggs, selection also ought to act on females to control that access. Differential fertilization of eggs by different males can arise from a number of mechanisms, some benefiting females, some benefiting males. Males benefit if they at least fertilize some offspring, even if they cannot fully remove the sperm of a female's other mates. Females may benefit if they can sequester some eggs to be fertilized by later males if a better mate comes along. Neither sex is likely to achieve a final victory, but we need to pay attention to both players.

CRYPTIC FEMALE CHOICE

Why should females attempt to influence which male's sperm they use? Two answers to this question have been proposed. The first is one of the same reasons that females prefer one male over another to begin with: some mates are better than others, either because they supply the female with better material benefits, such as protected nest sites or food for the offspring, or because they carry genes that will enable their offspring to resist disease or grow more efficiently than offspring with an inferior inheritance. The idea that females can detect such "good genes" by examining the secondary sexual characteristics, like the elaborate tail of a male peacock, is still somewhat controversial, and it is not clear how often such a process occurs. Nevertheless, in many species, including wolf spiders, fruit flies, crickets, and tiny spider relatives called pseudoscorpions, as well as a selection of vertebrates, females allowed to choose their mates have offspring with greater viability than do females mated at random to perfectly serviceable males they did not prefer. A logical extension of this suggestion, then, is that if a female has mated with more than one male, perhaps she can then select the sperm bearing the best genes to fertilize her eggs. Such a procedure might even have some advantages over exerting the preference behaviorally by surveying many potential mates but then only mating with one preferred male; females could stash away the sperm of several individuals and choose the best at their leisure rather than having to go back and forth among the males themselves. Whether females can in fact exert such control over fertilization remains to be seen, but we will never find out if we do not look in the first place.

The second answer to why females may choose from among males' sperm has been championed by biologist William Eberhard, who studies a variety of animals, mainly insects, in tropical America. Eberhard displayed early in his career a fascination with genitalia that most people would find unhealthy, if not downright pathological. Like Göran Arnqvist, discussed in the next chapter, he has documented the complexity of shapes and sizes of the reproductive structures in a great many species (a figure legend in one of his books rhapsodizes about the "intricate and beautiful genitalia of a male medfly," a phrase not frequently seen in print). He has gone further, and developed a theory of what is called "cryptic female choice," a term originated by Randy Thornhill to distinguish the obvious physical choice of a particular male by a female from the more subtle— cryptic—selection among males that may occur inside a female. Cryptic female choice cannot be detected by counting the number of copulations. Eberhard claims, in most detail in a book he titled *Female Control,* that sexual selection itself can be driven by the discrimination that females exert among the genitalia and sperm of males.

Unlike the good genes theory, Eberhard's idea is that selection for male traits that stimulate the female to produce fertilized eggs can occur even if the males that provide more stimulation aren't necessarily carrying better genes for viability. In a book chapter published after *Female Control,* he states, "In general terms, if a female's traits, whether they be behavioural, morphological, or physiological, have the effect of consistently favouring some conspecific males that have copulated with her and that possess a particular trait that in other mates is less fully developed, it is reasonable to conclude that the female traits bias postcopulatory male competition in a way that favours possession of that male trait, thus producing cryptic female choice" (p. 94). This rather dry statement is noteworthy because it expresses, in the most scholarly terms, what is ultimately almost a hedonistic view of female sexual behavior. The way Eberhard sees it, females "set the rules of the game."

For example, male katydids of many different species produce not only a small sperm-containing spermatophore like crickets, but a nutritive blob attached to it called a spermatophylax. The female eats the spermatophylax, and its protein-rich contents enable her to lay more and larger eggs. The sperm are transferred to the female while she is eating the spermatophylax; when she has finished her meal, she often reaches around, breaks off the sperm-containing structure, and eats that too. The larger the spermatophylax, the longer it takes her to finish it, and therefore the more sperm enter her body. According to Eberhard, "This arbitrary sequence of female behavior represents a 'rule of the game' for males. The female could as easily break off the spermatophylax, set it aside and eat the ampullae [the sperm-containing portion] with most of their sperm still inside. . . . By consistently imposing this and other, additional rules, the female . . . favors males which make a larger spermatophylax by laying larger eggs that he has fertilized. An increase in the tendency to discard the spermatophylax or to consume it faster could change the venue, imposing different selective pressures on males" (*Female Control,* p. 17).

The book lists an exhaustive series of other examples, many deriving from the complex genital structure of insects, documenting opportunities for females to influence the fate of sperm, ranging from folds in the walls of the sperm-storing organs that can hinder the removal of sperm by subsequent mates to biasing the transport of sperm from particular males. Eberhard accuses biologists studying reproductive behavior of "fertilization myopia," in which only the successful, "normal" fertilizations are noted, while the variation in how successful each mating attempt may be is ignored. He also points out that most laboratory studies use what he calls "easy" females, those that are in peak reproductive condition, often virgins, and generally young, because those females are predisposed to mating quickly and becoming pregnant. But that's precisely the problem: in nature, many females aren't at their peak. They aren't "easy." They may be malnourished, at the end of the breeding season, already paired with another male, or just plain inexplicably not interested. This difference can lead to incorrect conclusions about how reproductive behavior evolved, because it presumably did so in nature, not in the laboratory. How do we incorporate the way females really are into the way we study them?

Other researchers as well as Eberhard had noticed something interesting about those classic P2 values, too. The example I gave earlier made it seem as if all second or first males had the same share of paternity, depending on the pattern of sperm precedence in the species. But the P2 value can vary enormously; one study of a moth showed that sometimes the second male fertilized all the eggs, sometimes none of them, and sometimes about half. This variation used to be attributed solely to the different competitive abilities of the males involved; anyone can win a fight with a scrawny male, but most have more difficulty with a burly one. This interpretation again views the female as the passive arena. But could the female herself make a difference? Some researchers are starting to examine the fertilization abilities of the same pair of males when they mate with different females, and to at least consider that the female herself could be using sperm of different males selectively.

Once the possibility of cryptic choice and active females is raised, many more questions arise as well. What is the function of portions of the female reproductive tract? One scholarly paper on sperm competition in birds peevishly notes, "The vagina is a particularly hostile region of the female tract . . . ," which to my mind takes entirely the wrong starting point. Why is it called hostile, rather than selective? Hostile toward whom? If all those sperm got through, they'd still be out of luck when it came to insemination, since there are always a lot more of them than there are eggs.

Patty Gowaty wrote an article about a recent conference on female control of paternity, in which scientists from the United States and Europe presented reports of behavior in birds, fish, and mammals showing potential avenues for female actions to affect evolution. The title is "Architects of Sperm Competition" (perhaps a deliberate departure from those ubiquitous sports analogies, which she dislikes), and it is one of the harbingers of a new way of thinking about sperm competition, or what might more accurately be called sperm management.

What happens to sperm inside the bodies of females may turn out to have many implications for several important issues in evolutionary biology. First, the potentially variable fate of sperm puts a different perspective on forced copulations, and on aggression between males and females. If females can alter the likelihood of fertilization after sperm are deposited, even an apparent failure to resist mating may not be the end of the story. Sperm are stored in many species of animals for periods ranging from a few hours or days to many months or even years (for example, in social insects, where queens mate during brief flights after they become adult but never leave the colony thereafter). In itself this presents an opportunity for sperm management, but in addition some females actively destroy sperm. This discovery comes from a few species of salamander, but it is not yet known which sperm are destroyed or why.

Second, sperm management affects another new area in sexual selection: sexual conflict. Although evolutionary biologists have recognized for a long time that male and female interests are different, this subject has received a new slant lately. Erroneously assuming that the sexes cooperate for the "good of the species" is likely to neglect the female perspective in particular, because the male often serves as the model, the norm. Suffice it to say that if females can actively deal with sperm, they are continuing to resist male efforts to control paternity just as males continue to compete after ejaculates are placed in the female tract.

Finally, because sperm competition or management presupposes that females mate with more than one male, studying the fate of sperm from those males means studying the nature of relationships. Why should monogamy occur, if females benefit from multiple mating? What are the risks and benefits of mating with two males? With three males? With three dozen? How can sperm from different males be distinguished once it is in a female's reproductive tract? As in the study of extra-pair copulation, once it becomes clear that what you see is not always all that is happening, and that what is happening changes if you take on a female perspective, it becomes possible to think about mating systems, and therefore male-female relationships, in a completely new way.

PART TWO

Unnatural Myths

Six

SEX AND THE *SCALA NATURAE*
(OR, WORMS IN THE GUTTER)

WHAT IS OUR PLACE IN NATURE, anyway? A recent lecture on the evolutionary basis of human mate choice at my university was titled "Sex, Evolution, and Dynamical Systems: Lying in the Gutter Looking Up at the Stars." The catchy subtitle comes from Oscar Wilde, who said, "We are all in the gutter, but some of us are looking at the stars." The lecturer used this image in claiming that "though searching around in the gutter, examining sex and aggression and comparing humans to dogs and baboons, evolutionary psychologists keep one eye on the stars—working towards an integrated conceptual paradigm to unite psychology's scattered subdisciplines and neighboring sciences." The implicit idea is that although we humans are mucking about in the gutter, nonhuman animals are even "lower" and are not capable of looking at the heavens with our sophisticated understanding.

This is an old, and, to some, not even a particularly controversial concept, this notion that humans are higher, more advanced, more evolved, less primitive, or (if you get right down to it) better than other forms of life, which wallow around in literal or figurative gutters while we humans contemplate astronomy, astrology, or theology, whichever form of looking at the stars seems most suitable. The American philosopher and poet Ralph Waldo Emerson put it another way:

> Striving to be man, the worm
> Mounts through all the spires of form.

The division between humans and other animals is not generally seen as a simple dichotomy, either. One may view humans as unique either as a result of practical reasoning (after all, we are the ones with personal computers, nuclear weapons, and literature) or because of one's philosophical ideas. But beyond this initial division, most of my students and more of my colleagues than I like to admit also subscribe to Emerson's view that animals can be ranked, with worms (or something similar) at the bottom, and humans (not coincidentally usually called "man") at the top. It is as if we were all in a planet-wide military, with people as generals and protozoa as privates. My students in particular are convinced that taxonomists and systematists, the biologists concerned with classification of organisms, spend their time busily placing each and every species on what is often termed an evolutionary ladder. Each species is assigned to a rung, and each rung has many stretching above it, until we get to people, who are sitting at the top. Learning zoology means learning a sequence, so that they must memorize the position of animals relative to one another. Anatomy textbooks seem to have bought into this notion with particular zeal, and frequently present the sequence of dogfish (*Squalus acanthias,* a type of shark) → frog → cat → human as if this were how evolution had occurred. The same kind of thinking has pervaded animal psychology, with animals supposedly better able to learn and accomplish sophisticated behavioral tasks, like monkeys, seen as being a few rungs away from humans rather than many rungs away, like goldfish or flatworms.

What is wrong with this idea? Why shouldn't we look at life as a gradual progression of forms, ever increasing in complexity, with each organism paving the way for those that come after it? In this chapter I trace the history and use of this idea of a *scala naturae,* or scale of nature, and suggest that it is a myth that has led us, particularly with regard to sex roles and the evolution of gender differences, quite seriously astray. We can, I believe, still learn a great deal by comparing the behavior of different kinds of animals, but we need to use different tools, and start out with different assumptions, from the ones we may be used to. (I will not be able to cover the related concept of progress or major transitions in evolution, but the interested reader is referred to works by John Maynard Smith and Eörs Szathmary.)

SCALES, LADDERS, AND ARISTOTLE

In a classic paper published in 1969 and boldly titled "*Scala naturae:* Why There Is No Theory in Comparative Psychology," William Hodos and

C. B. G. Campbell point out that the idea of a hierarchical classification scheme for organisms is not new. Aristotle attempted to organize the world's plants and animals, not only according to characteristics such as number of legs or whether an animal was aquatic or terrestrial, but along a unitary scale called the *scala naturae* or Great Chain of Being. This scale had man (I use the word advisedly) at its apex, and complexity was equated with progress toward perfection. In a 1936 book titled *The Great Chain of Being,* A. O. Lovejoy connected this idea with the more spiritual—and also ancient—notion that all creatures are to a greater or lesser extent copies of God. "Thus," write Hodos and Campbell, "angels were somewhat imperfect copies, man more imperfect, apes still more imperfect, and so on, down the scale to the 'formless' sponges" (p. 247).

This theological attitude, Hodos and Campbell go on to point out, transmogrified itself into an apparently respectable biological concept, that of an evolutionary or phylogenetic scale. Hence the reliance in textbooks on an organization that first examines a "lower" animal, whether this is an ameba, an earthworm, or a fish, and then proceeds to "higher" forms of life, again depending on the topic, but perhaps a frog, reptile, and finally mammal. Within the mammals, primates are usually seen as "higher-ranking" than, say, rodents. Birds have an uneasy position in the picture because it's not clear whether they are "above" or "below" the mammals; they are descended from reptiles, as are the mammals, but clearly did not give rise to mammals themselves. Such difficulties aside, mentions of a "phyletic scale," an "evolutionary scale," or some similar entity abound in both written and spoken discussions of biotic diversity. Linnaeus, the originator of the scientific naming system still in use, also saw relationships in terms of a chain of being, with all organisms representing their creator's design.

It is not only my students and their textbooks that assume the existence of a *scala naturae.* In my research on sexual behavior in red jungle fowl *(Gallus gallus),* birds that are the ancestors of domestic chickens, I must deal with the campus committee that oversees animal research. This committee ensures that I comply with national regulations on the care and use of animals, and every institution in the United States that relies on federal funding for research has its equivalent. I am required to fill out forms, file reports, and answer questions about the ways I perform experiments and make observations. The funny thing is, if I worked on worms, or starfish, or butterflies, or sea anemones, I would not need to do any of it—because, according to this committee and others like it, none of those organisms are considered animals. What are they, you may ask? Who knows? Most of us simply roll our eyes and thank our lucky stars that some forms of research require a little less paperwork. We do know, however, that the reason for the regulations being the way they are is the belief in a *scala naturae*. It matters more to people if you work on animals that they somehow see as being closer to humans. But closeness is still defined in an oddly Aristotelian way. Some of the bias is freely admitted to be subjective (rabbits are cuter than mice and monkeys are cutest of all, whereas birds and reptiles don't inspire that same cuddly feeling), but it rests ultimately on a belief that we need to be more concerned when animals are closer to us on this evolutionary ladder. If they are far enough away, we can even get away with not calling them animals, at least for some purposes.

The psychologist in the example with which I began this chapter suggested that he would compare human behavior to that of dogs and baboons. Why dogs and baboons, and not grasshoppers or jellyfish? Because we assume that creatures that are higher on the *scala naturae* must be more like us, even though the reason we put them in their position on the *scala naturae* in the first place is that we thought they were like us to begin with. The behavior of primates or other large mammals somehow is thought to have more validity in explaining our own behavior than that of the "lower" animals. It is hard to imagine Wing Bamboo, the would-be sibling of the loon, saying "Goodbye, Brother Earwig," but does that mean that earwigs are less interesting and valuable only to New Age believers, or also to scientists? Donna Haraway wrote a book about scientists, scientific practice and gender called *Primate Visions: Gender, Race, and Nature in the World of Modern Science*. It is hard to imagine her having titled it *Nematode Visions* instead, even though studies of nematode worms are at least as common in science as studies of primates, and the scientists studying them at least as diverse as primatologists.

A THUMB ON THE *SCALA*

Well, so what? Surely evolutionary biologists agree that some animals are more closely related than others, and that mammals like the dogs and baboons arose more recently in evolutionary history than did the jellyfish or grasshoppers. Since humans, too, arose more recently, shouldn't we look to other similar species for clues about the origins of our own behavior? And furthermore, what does this have to do with sex and gender roles?

Quite a lot. First, the *scala naturae* simply does not exist. It is confused, frequently, with a phylogenetic tree, shown in Figure 1. A phylogeny is an evolutionary history that uses the branches of a tree to indicate divergence from historical ancestors. Time moves vertically, so that the tips of the branches are all species that exist now, and their ancestors occurred lower down, or longer ago, on the tree. Thus in the diagram, species A, B, and C all have a more recent common ancestor among themselves than existed between any one of them and species D, and furthermore species A and B have a more recent common ancestor, or are more recently derived, as it is called, than either has with any of the other species.

Historical relationships exist among all forms of life, with earlier groups, such as mastodons, giving rise to more modern ones, such as elephants. Through evolutionary time, genes producing mastodonlike proteins changed their frequencies in the population, the population diverged, and eventually genes producing more elephantlike proteins became more common. Mastodons are the ancestors of elephants, but elephants are no "higher" or "lower" than mastodons, simply more recently evolved. The tips of the branches on the tree represent more or less distinct groups of genes that we can differentiate into species, not a continuation of life into more and more sophisticated forms. I would be the last to argue that the White Leghorn is closer to humans, much less to God, than the jungle fowl I use for my research, although the familiar barnyard breed indisputably arose more recently. Yet the blind spots that the myth of the *scala naturae* induces infect many of our perceptions of what animals are and do.

The notions of "lower" and "higher" are themselves highly disputable. Most people would have a difficult time ranking, for example, snails vs. starfish, or hamsters vs. mink. Known biodiversity simply was not as immense when Aristotle and his devotees tried to develop their ladder. And even the gross relative ranking of fish and amphibians falls apart when certain characteristics are examined; many bony fishes (as opposed to the cartilaginous ones like sharks and rays) have more complex central nervous systems than your average salamander, despite the latter's emergence onto land, a hallmark of advancement to which many textbooks pay lavish tribute.

The concept of recent evolution is a tricky one in other ways as well. Strictly speaking, with the textbook definition of evolution as a change in gene frequencies in a population, many of the most rapidly evolving species, and hence those with the most recent changes, are not primates but pathogens, the disease-causing organisms like viruses and bacteria. Because

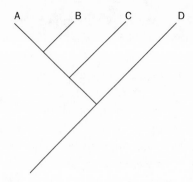

FIGURE 1. *A sample phylogenetic tree showing the evolutionary histories of species A, B, C, and D. Time is on the vertical axis, with the present at the top of the figure.*

of their rapid generation times, viral gene frequencies can become altered in a fraction of the time it would take to do the same thing in a population of humans, zebras, or any other vertebrate. The human immunodeficiency virus that causes AIDS, then, is much more recently evolved than we, its hosts, but there is not a lot of hue and cry to change its position on the evolutionary ladder and place us below HIV or the bacteria that causes tuberculosis. The crocodile looks rather unchanged from its days in the primordial soup, and television shows and texts often refer to crocodiles rather condescendingly as "living fossils," as if they were stuck in the past, wearing the evolutionary equivalent of bell bottom trousers. From the point of view of a syphilis bacterium, however, the human form is just as hopelessly dated. Not only do organisms not lie in a neat stack of ever-increasing complexity; humans do not occupy the top of that stack, and never have. If worms want to strive for anything, they might consider striving to attain the level of viruses or bacteria.

Yet because the *scala naturae* encourages a relentlessly hierarchical way of thinking about organisms, in addition to trying to rank groups such as amphibians relative to reptiles or reptiles relative to mammals, taking it to its logical extension, one tries to rank each species. But why stop there? Many attempts have been made to rank the various races of humans, with Western Caucasians placed in a higher position than people of more recent African ancestry (we are all, of course, of African ancestry). Similarly, women have been viewed as less highly evolved than men, because the female reproductive tract is thought to make us more animal, less cerebral, and therefore somehow less complex or sophisticated. How finely can we draw distinctions here, and who gets a rung of the ladder to him- or herself?

I do not mean to be disingenuous and argue that every species of plant, animal, or microorganism is just as complicated as every other, or that it makes just as much sense to compare humans to worms as it does to compare them to chimpanzees. Certainly if one wants to see, for instance, how the vertebrate kidney may have evolved it is a better idea to compare animals that possess kidneys than those without them; making use of the known evolutionary relationships is appropriate, and there is no point wasting time with spiders rather than other vertebrates. But what about studying behavior, such as the origin of monogamy, or the prevalence of female rather than male parental care?

We have grown to rely on certain species for our most meaningful lessons about behavior: primates, social carnivores like wolves, a few other group-living mammals such as dolphins, and songbirds. The discovery of tool use by chimpanzees rocked the world, even though a variety of wasps perform feats with sticks and stones that put the chimp grubbing for termites with a grass stalk to shame. We have known for many decades that butterflies, for example, mate with many sexual partners, some species of cockroaches are monogamous, male seahorses get pregnant, and that for many types of fish, sex roles are completely blurred within individuals because they either change sex or have the sex organs of both male and female contained in the same individual. Somehow, though, these discoveries are not as meaningful for us as those made about primates or wolves. When female macaques, the monkeys used in medical research, turn out to actively seek matings from a variety of males, it disturbs the stereotype of the passive female. When male primates in a tiny minority of species, such as marmosets, are found to help take care of their offspring, it is heartening news for fatherhood and the New Age male. If we see an animal as being closer to humans on the mythical *scala naturae,* we, unconsciously perhaps, weight evidence of its virtues and vices more heavily than if we discover the same behavior in a more distant relation. Sexually active butterflies do not do much to shore up the feminist argument that females are not merely passive sex partners.

There are two problems with this stacking of evidence. The first is that we should not be using animals to guide our ethics in the first place, regardless of which side of the social or political fence we prefer. The second is that no a priori reason exists to pay more attention to monkeys than moths, if all we are doing is exploring diversity in animal mating behavior to see what patterns may exist and what selective pressures may have led to them. It is no more instructive to find that males do all of the parental care in pipefishes than it would be to discover doting fathers among baboons, and yet the absence of such paternal care in many primates has led many people to conclude that human mothers are "biologically" more prone to care for children than are fathers.

Still, one might say, if other primates are more recently diverged from our common ancestor than pipefish, surely this means that we all inherited similar tendencies to behave in a certain way, doesn't it? Can't we still extrapolate from what other animals do to humans? The answer is both yes and no. If we were concerned with whether female monkeys mate with more than one male only because we were concerned that submissive and sexually uninterested females were being touted as the norm, and if we assumed that what female moths did was irrelevant because they were so far "below" us, we would have missed an important and interesting general pattern among animals. The absence of a *scala naturae* is, I think, liberating. It allows us to incorporate more model systems, because we are no longer constrained by models that resemble us in morphological—and therefore presumed psychological—detail. Although we expect to be roughly similar to our relatives, we are also similar to other creatures, and we need not be slaves to a few primate species regardless of how feminist or sexist we might find their behavior.

YANKING AWAY THE LADDER

Two recent studies of sex in animals illuminate how evolutionary relationships without a *scala naturae* can teach us about both the animals and ourselves. The first is an (as yet unpublished) examination of the evolution of so-called reversed sex roles in wading birds, from two Swedish scientists, Jacob Höglund and Jobs Karl Larsson. Most birds show what many people might call conventional sex roles, so that in typical 1950s fashion, mom does more of the parental care and dad is mainly interested in courtship and conception. Höglund and Larsson wanted to explore some of the exceptions to this common division of labor. These exceptions occur in several species of shorebirds including the three types of phalaropes (Wilson's phalarope, *Phalaropus tricolor;* the red-necked phalarope, *P. lobatu;* and the grey phalarope, *P. fulicarius*) and the spotted and common sandpipers *(Actitis macularia* and *A. hypoleucos),* all long-legged inhabitants of

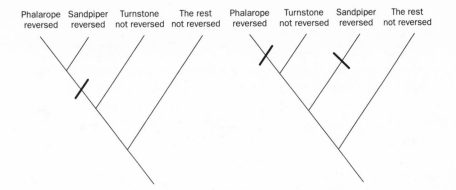

FIGURE 2. *Two scenarios for the evolutionary development of reversed sex-role behavior in phalaropes and sandpipers.*

coastlines throughout the world. In all these birds, females are larger and more colorful than males, and a female may mate with several males in succession, leaving a clutch of eggs with each, so that the father cares for the chicks that hatch while females are free to compete for further matings.

The question is whether this behavior arose independently more than once in evolution, which would suggest that male and female roles are quite flexible and can respond to environmental pressures by evolving through natural selection. Alternatively, the reversal is a fluke, an oddity that arose once in an ancestor of all of these species and simply stayed with them, to be shared the way that they share a long bill for probing the sand for small worms and crabs. To illustrate, does the phylogenetic tree for these birds look like the left portion of Figure 2, in which the ancestor of both phalaropes and sandpipers evolved reversed sex roles and its descendants retained it? Or did the behavior arise separately, through different evolutionary events, in the two genera, as shown in the hypothetical tree on the right? In both cases, the turnstones *(Arenaria)*, a genus of two species of similar shorebird, but with conventional sex roles, are shown on the trees to contrast with the role-reversed species. The thick black lines indicate the time at which the trait of role reversal evolved in the two scenarios.

At first glance, this might seem to be an entirely hypothetical question. After all, no one can go back in time to see what the ancestors of sandpipers were doing in their family lives several million years ago. But modern technology and analysis of DNA give us the possibility of an answer. The similarity of genetic material in different species is directly proportional to the amount of time since the species diverged from each other; chickens and sandpipers have fewer genetic sequences in common than do chickens and turkeys, for example. Höglund and Larsson obtained samples of DNA from all the species of birds they wanted to study, including some from what are called outgroups, or species that one does not expect to share an unusually large amount of genetic material with the groups of interest. The outgroups serve as a kind of calibration for the technique. Several different methods are used for this kind of analysis, with each having its proponents and detractors. Höglund and Larsson used one of the more common techniques, the sequence of proteins in a gene called cytochrome b, and came up with the results diagrammed in Figure 3.

At the tip of each of the branches is a species of shorebird, with the occurrence of reversed sex roles indicated for two distinct groups, one on the right and one on the left. It turns out that the phalaropes, shown in the three twigs on the far left, have as their closest evolutionary relatives not the other role-reversed species, but three other kinds of sandpiper: the greenshank *(Tringa nebularia)*, the greater yellowlegs *(Tringa melanoleuca)*, and the black-tailed godwit *(Limosa limosa)*. The role-reversed common and spotted sandpipers *(Actitis hypoleucos and A. macularia)*, on the far right, are more genetically similar to the turnstone *(Arenaria interpres)* and the great snipe *(Gallinago media)* than they are to the phalaropes, but the turnstone and snipe have conventional sex roles. Hence, concluded Höglund and Larsson, the pattern of males taking care of the offspring and females competing with other females is what is termed paraphyletic: it arose independently more than once in a lineage of even the quite closely related (in the greater scheme of things) shorebirds. Just why female turnstones stay home with the chicks and female common sandpipers do not is an interesting question, but one that Höglund and Larsson's study cannot address. The point is that even within a small group of similar species, which sex does what is far from being rigidly determined, and a Donna Reed–like maternal role is no more or less "natural" than any other.

Now, even if this study had shown that sex role reversal originated only once, I would hardly argue that we should meekly conclude that women are supposed to stay at home and take care of their children while men act as breadwinners out in the working world. As I have pointed out repeatedly, I do not believe we should be using animal behavior to direct our own. We can, however, draw two conclusions from Höglund and

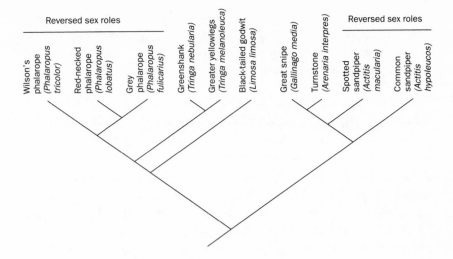

FIGURE 3. *Evidence of multiple origins of sex role reversal in shorebirds, suggesting that sex-typical behavior is evolutionarily flexible.*

Larsson's work. First, sexual behavior in animals is extremely flexible, and even such a basic characteristic as which sex takes care of the offspring is not always the same in species that otherwise differ only in a few crumbs of amino acid sequence. Parental care patterns can differ even within a species, often depending on environmental circumstances, such as whether the food for feeding the offspring is available over only a short period, or can be acquired at a more leisurely pace, over many weeks or even months. Second, in saying that we should abandon the myth of the *scala naturae* I am not suggesting we abandon comparing animals to each other. Comparison does not mean modeling, as I discussed with regard to model systems. We need not even abandon our affinities, emotional and physical, to many primates; we simply need to recognize that they do not necessarily mirror our selves. It is, however, much more meaningful to try and understand evolution of behavior—or any other trait—using actual phylogenetic relationships, rather than a trumped-up Chain of Being with no basis in biological reality.

Another example of the exciting and appropriate use of evolutionary information to compare animals comes from an even more esoteric area than the sex lives of shorebirds. Most of us, I would venture to say, take the general size and shape of our genitalia for granted. We pretty much assume that the pattern of "extended male part inserts into hollow female part and enables sperm to meet egg" is the way it has to be, and any individual variations are the stuff of anxious locker room comparisons, adolescent paranoia, and the workings of the minds of pornography writers.

It turns out, however, that many animals have baroque constructions of male and female parts that resemble bits of the decorations on Old World cathedrals. Little knobs and ridges on the intromittent organ of males (strictly speaking, it is called a penis only in certain groups of animals) match grooves and funnels in the interior of females of the same species, and even closely related species may have quite divergent shapes and sizes of genital structure. This elaboration is especially pronounced in many insects, although some mammals have their fair share of kinky anatomical devices for transmitting and receiving sperm. In many species, the shape and structures of the male genitalia are the only reliable characteristic for separating closely related groups.

A study published by another Swedish scientist, Göran Arnqvist, in the prestigious scientific journal *Nature* uses those genitalia to provide a fine example of how evolutionary relationships can shed light on sexual behavior. Arnqvist pointed out that two competing hypotheses could explain the pattern of elaboration. First, the so-called lock and key hypothesis, which taxonomists have long favored, holds that male and female genitalia are very complicated in certain species because they fit together like—you guessed it—a lock and key, and only a male of the correct species can unlock, so to speak, the portals leading to the castle tower wherein lie fertilization and evolutionary success. The point is that the necessity for a close congruence between male and female parts renders hybrids between two species unlikely, which is usually useful from an evolutionary standpoint because such hybrids, if they can be produced at all, are generally less fit than either parental type. Mules, which result from such a hybridization between horses and donkeys, are always sterile, which is of course an undesirable outcome because the genes of both parents are at an evolutionary dead end.

The second hypothesis is newer, and takes into account the competition that can occur both between different males that mate with the same female and between males and females over control of fertilization (the ultimate in playing hard to get). This second idea, called the sexual selection hypothesis, suggests that what governs the evolution of complex and divergent genitalia is the likelihood that a female will mate with more than one male in a given reproductive event. If a female mates with only one male, competition for fertilization of the egg is likely to be weak, whereas if she mates with several, ample opportunity exists for selection of one of them. The competition among males and the selection of one from an assortment should therefore select for ever-increasing elaboration of the mechanism for achieving fertilization, the genitalia. The lock and key hypothesis, in contrast, suggests that mating with only one male should select for more intricate genitalia, because it is extremely important in such cases to avoid the peril of hybridization, given that the female has only one chance at making the correct connection.

Arnqvist examined many different types of insects to see which of the hypotheses was supported by the data. The prediction is that in species where females mate once, genitalia should be more elaborate than in multiply-mating species if the lock and key hypothesis is correct, but less elaborate if the sexual selection hypothesis is correct. He obtained the information both on genital structure and on mating patterns in nineteen different groups of butterflies, mayflies, beetles, and flies by finding articles published in the scientific literature and by contacting entomologist colleagues on the telephone or via email. (It is not as unexpected as you might think to receive a phone call from someone asking you to describe the mating habits and genitalia of the group of insects you work on.) The methods for determining what constitutes "more elaborate" in the kinks and whorls of the seminal apparatus of a moth need not concern us here. It was also important to control for the effect of evolutionary history, as with the shorebirds, so knowing the phylogenetic relationships is again crucial. After gathering all the evidence, Arnqvist came to a main conclusion that was simple and stunning: "Genital evolution is more than twice as divergent in groups in which females mate several times than in groups in which females mate only once." Sexual selection, not the avoidance of hybrids, seems to have influenced the way in which one of the most important, perhaps *the* most important, components of animals has evolved.

What does this mean for our views about gender and biology? As with the previous example, there is no real agenda being supported by Arnqvist's results, feminist or otherwise. It would be wrong-headed to have started either this research or the shorebird project with the aim of showing that females rule the world, or even that feminism should direct our hypotheses. I am generally pleased by the idea that behavior of females shapes the pathway of evolution, and there is evidence for that point of view, but it is incidental to the findings. As for the *scala naturae,* it is equally incorrect whether it is used to buttress or defeat sexist ideas.

In a larger sense, the more modern and enlightened use of evolutionary relationships among animals illustrated in my two examples shows that our understanding of behavior can be much richer if we do not limit ourselves to looking to "higher" animals in a few groups like primates. When scientists look at animals and try to generalize about behavior, we are often accused of mindless extrapolation, of saying that what is sauce for the goose is sauce for the woman. It is a legitimate concern: If other animals are like us, does that mean we are limited by what we see them do? If, however, they are not like us, then are we stuck being unable to learn about ourselves from other organisms?

Removal of the *scala naturae*, however, gives us a way out of the dilemma. What we are looking for is the product of natural or sexual selection, of how organisms respond to particular environmental pressures in evolutionary time. Abandoning the *scala naturae* means that the selection becomes the focus, not the animal, so that it is possible for us to be like pipefish, or butterflies, or gorillas, in the sense that we see how similar selection can yield similar outcomes. If intense male competition tends to favor larger individuals, because those individuals fight better and end up siring more offspring inheriting the genes for being large, we might expect to see beefy males in many species, vertebrate and invertebrate, regardless of how recently they had a common ancestor. At the same time, because behavior is an interaction between genes and the environment, we cannot predict with certainty how a given species will respond to selection; we can only suggest a probable outcome to be subject to further testing. Thus, if male rhinoceroses are larger than females, we might predict that they engage in competition over females. Careful observation might or might not substantiate this claim; we have simply used our understanding of how selection works to generate a likely hypothesis. Although we used information about other animals, some closely related to the rhino and some distant, to help us, no one is suggesting that rhinos are always like these other species, or that they are like them in every particular. In the case of the shorebirds, role reversal can arise from the same pressures at different times. In a similar vein, abandoning the *scala naturae* also means that we understand that even very closely related species may respond differently to selection, or may be subject to entirely different forms of selection and other forms of evolution, such as the rate at which mutations appear. If there is no ladder, then we do not need to worry about who is at the top of it, and we can learn as much from Emerson's worm as from his more apelike relatives.

Seven

BONOBOS

Dolphins of the New Millennium

OUR PREDILECTION FOR RANKING animals in terms of their similarity to us, for looking at them seeking reflection and maybe affirmation of our own behavior, for turning some into role models, not only prevents us from seeing what the animals are actually doing. It also leads us to use some species as mythical figures, forms that serve as symbols and totems. An animal with which we identify can come to stand in for our values and desires, and then to validate the mysteries of nature to us in return. Some of us are "cat people," some "dog people." We want to save species from extinction, but let's face it, we want to save some more than others. What is good and bad about this kind of identification?

Remember "Save the Whales"? Remember Flipper, and freeing Willy? For a time, cetaceans, the animal group that contains dolphins, porpoises, and whales, seemed to exemplify our fascination with nature and the growing environmental movement. Dolphins were hip, they were free, they were as intelligent as we were but had somehow managed to escape mortgages, sex therapists, and cell phones. We had to preserve them, not because they were a vital part of the ocean ecosystem, but because they talked, at least to each other and maybe, if we worked on it hard enough and freed our minds from preconceptions, to us. Many people were convinced that communication between species held the key to a new understanding of ourselves. To some, whaling became almost as much an anathema as cannibalism. A dolphin's brain size relative to its body weight is impressively large, on the same order as our own, a finding that seemed to set the cetaceans apart from merely cute animals like dogs or squirrels. Whales and dolphins were unique, different from pets, certainly not to be domesticated like cows and chickens. Placing them in aquariums to do tricks and wear funny hats bothered some people, who thought the animal acts were demeaning to creatures of such intelligence.

The dolphins, however, are no longer quite such animal icons. The fascination with whales and dolphins has ebbed, to be replaced, it appears, with a love for another species with a potential for complex communication: the bonobo, a relative of the chimpanzee with the added advantage of a naturally kinky sex life. Visitors to zoos with an exhibit of these small apes often emerge red-faced from their encounter with bonobos; the animals have a very overt sexuality which is expressed in interactions between individuals of all ages and both sexes. Pairs of females often engage in a behavior called "genital-genital rubbing," or GG-rubbing for short, which is exactly what it sounds like. Groups of bonobos have been studied over the last two decades both in captivity and in their native Africa. In addition, they have proved to be willing students in the "ape language" trials, attempts to teach primates to use symbols to communicate with humans. In their own social system, sex appears to have replaced aggression as a way of dealing with conflict; almost everyone who was alive in the 1960s and a few who were not refer to bonobos as the original proponents of "make love, not war."

This high intelligence and apparent lack of sexual inhibitions have made bonobos the latest animal celebrities. In a web site devoted to promoting bonobo sexuality as a model for our own, phone sex therapist (don't ask) Dr. Susan Block rhapsodizes, "Like tantric sex practitioners, or just like two people very much in love, copulating bonobos often look deeply into each other's eyes." Bonobos, she asserts, are "using sex to make peace." She explicitly advocates what she terms The Bonobo Way, which even she admits is "a very simple philosophy (after all, these aren't geniuses, they're chimpanzees) . . . Pleasure Eases Pain, Good Sex Defuses Tension, Love Lessens Violence." If we leave aside the question of whether we needed to send scientists to Zaire to watch apes having sex in order to come up with these insights, Block is only doing what people have done for centuries: taking animal behavior and using it to champion a viewpoint. That she appears on television wearing fishnet stockings and carrying a velour vulva puppet just accessorizes her stand a little differently from most.

The juxtaposition of dolphins and bonobos is intriguing for several reasons. Why did we embrace and then abandon the dolphins, and what is it about the bonobos that made them good replacements, or at least viable co-stars? In this chapter I explore what it means to use these particular animals as icons, as near-mythic symbols, and why we show our biophilia, or love of the natural, so selectively. We love the cetaceans for their minds, and the primates for their bodies, but we risk losing the animals themselves in our zeal to express that affection. In the case of the bonobos, using them as feminist symbols can blind us to their inherent value.

WHALES, DOLPHINS, AND THE MONA LISA

Humans have felt a special connection with cetaceans for thousands of years. Anecdotes of dolphins that apparently voluntarily sought out interactions with people have been popular at least since the ancient Greeks, who also recognized that the animals were not fish but air-breathing mammals like themselves. Some of these anecdotes suggest that dolphins rescued drowning humans, pushing them to shore. Images of dolphins appear in ancient art, and both McIntyre's *Mind in the Waters* ("a book to celebrate the consciousness of whales and dolphins") and Devine and Clark's *The Dolphin Smile* contain essays, poetry, and drawings from sources as diverse as Melville, Wordsworth, Pablo Neruda, and various Native American traditions. Because of the anatomy of the dolphin skull, the dolphin smile, like that of the Mona Lisa, is endearing, slightly enigmatic, and involuntary.

Modern fascination with dolphins owes a great deal to the efforts of John C. Lilly, a medical doctor who became convinced in the 1950s that dolphins were unusually intelligent, that they had advanced methods of communicating among themselves, and that it was not unreasonable—in fact, it was pioneering—to attempt to speak with them himself. In *Man and Dolphin,* published in 1961, Lilly claimed, "Within the next decade or two the human species will establish communication with another species," and he was convinced that dolphins and perhaps their relatives the pilot whales would be those species. The dolphins were well known from animal acts in oceanaria such as Marineland and the Miami Seaquarium, where their trainers often told of pupils whose deductive capacities were startling and who seemed to perform not because they were given fish, but out of a fondness for interacting with their instructors. They are master imitators, and in a way that certainly suggests a comprehension of their activities; for example, a pair of female roughtoothed dolphins *(Steno bredanensis)* at Sea Life Park in Hawaii had seen each other perform different routines but had never been taught to copy each other. When the trainers inadvertently switched the performers, each dolphin did her companion's act so well it was only later that the mistake was noticed.

Lilly also was intrigued by the brain size of dolphins. The ratio of brain size to body weight is thought to be correlated with intelligence, perhaps because this ratio is relatively high in humans and apes. In some cetaceans, the ratio is actually higher than in most nonhuman primates. He therefore reasoned that these animals must be capable of a form of reasoning unknown to most species, and that it should be possible to reach across interspecies boundaries either to teach dolphins to produce sounds recognizable as human language or else to communicate in an as-yet-undetermined way. Anthropomorphism, he suggested, was not a problem because one needed to avoid attaching human characteristics to animals only "as long as the brain is very much smaller than ours." With cetaceans, all bets were off. The conservation connection, however, took on a slightly disturbing note; a species needed to deserve its preservation, and discussions about why we should preserve cetaceans centered not on the inherent value of biodiversity but on the particular fetching attributes of whales and dolphins.

These ideas caught on with a public starting to expand their own consciousness with sex, drugs, and rock and roll. The idea that an alternative intelligence, a consciousness completely unlike our own and yet able to be touched if the will was great enough, sparked the imaginations of countless people. It was as if extraterrestrials had already been discovered and we could be captivated by the possibilities while being assured that they were not going to take over our world. For one thing, they had no arms, which meant that they did not manipulate their environment. This made the animals less likely to launch a nuclear attack, but also profoundly influenced how they interacted with their world. Much was made of the differences that must exist between the dolphins' three-dimensional world of water, where sound travels great distances but objects are difficult to move, and our own. Dolphins gave us the astounding possibility that creatures completely unlike us physically might be very much like us mentally.

The icing on the cake was that many cetaceans, even the larger whales, seemed remarkably tolerant of humans, and some even actively sought contact with people. The tales of dolphins rescuing swimmers in distress are the most striking examples. Although as animal behaviorist Karen Pryor points out, the drowning people pushed *away* from shore are unavailable to give their testimony to the contrary, at least some of the claims of rescue have been reasonably well substantiated, and many scientists agree that cetaceans show a remarkable zest for interacting with humans that is not seen in most other animals either captive or wild.

All these pieces of information made dolphins and whales very popular, and the popularity not only drove T-shirt sales, it almost certainly helped pass conservation legislation protecting marine mammals from hunting and harassment. The Marine Mammal Protection Act was passed by the United States Congress in 1972. The act makes it illegal for anyone to kill, injure, or bother any and all species of marine mammals; these animals include not only whales and dolphins but seals, sea otters, manatees, and polar bears. It also makes it illegal to import marine mammals or related products, such as sealskin coats, into the United States. Its regulations—and the definition of "harassment" of wildlife, including the distances ships must keep from an animal in the water—are so stringent that a few scientists, including the famed whale researcher Roger Payne, worry that potentially valuable tourism or even worthwhile research will be unnecessarily curtailed. Countless private organizations support whale-watching trips that emphasize conservation, promote strict regulation of whaling, and sponsor "swims" with wild dolphins in protected sites all over the world.

Much of this fallout is undeniably beneficial for the cetaceans, and in many cases for other animals and marine life in general as well. But the origin of the fad is nevertheless troublesome, because it implies that dolphins are worthy of all this attention for the very reason that they are not like other animals, and furthermore because they are unlike them in a way we happen to find satisfying. Lilly himself left conventional science to, in the words of Joan McIntyre, editor of *Mind in the Waters,* "investigate his own mind, on the theory that the study of the self and the universe are one. His decision to concentrate on himself was prompted by the dolphins who, he feels, taught him a lot about being a human" (p. 71).

WINNING THE CONGENIALITY CONTEST

I have nothing against dolphins and whales teaching us about being human. But I am much more interested in having them teach us about being dolphins and whales, and I wonder why only certain animals, identified with characteristics they may or may not actually possess, are allowed to give us those lessons. Furthermore, I am worried that animals are being subjected to entrance examinations before they can be deemed worthy of either interest from the public or conservation of their natural populations.

A now somewhat outmoded idea in ecology is that of the balance of nature, in which each species was important and necessary because of its interaction with other species. Flies existed to provide food for frogs; frogs existed to eat flies, and to provide food for herons. Herons existed—you get the idea. The *scala naturae* rears its head here, too, but with emphasis not on the hierarchy but on the links among the levels, so that the Great Chain of Being becomes similar to a chain of paper dolls holding hands. If one clasp fractured, the whole disintegrated. A more sophisticated version is explained eloquently by Paul and Anne Ehrlich in their book *Extinction.* They liken the extinction of species to the removal of rivets from an airplane. Removing one rivet will not make the plane fail, but eliminate enough and the whole machine is in serious danger of collapsing. Similarly, ecosystems falter and topple if their members are destroyed, and one reason we should admire animals and want to preserve them is that they each serve an essential function.

The problem is that ecologists have been casting doubt on this simplistic interpretation of nature for some time. It is unarguable that organisms interact with one another, and that tampering with an environment like a lake or an alpine meadow can often have unforeseen consequences. But it would be rash to claim that each and every species fulfills an essential role, or that nature is so precise that every interaction is orchestrated.

My students often ask me, when I describe a species of animal they have not heard of, like a meerkat or a digger wasp or a comb jelly, what the animal "does." It took me a long time to figure out what they meant, since I assumed this was not the usual cocktail party query with an expected answer of "investment banker" or "pediatric nurse." Finally I realized that they wanted a phrase or sentence summarizing not so much the animal's use to humans as its role, its service to an ecosystem that they envisioned as operating like a large, somewhat socialistic, corporation, but with less waste and potential for corruption. Everyone has a task, and everyone is necessary, according to this view. Some species, then, are scavengers, and clean up the leftovers from other animals; some are the fastest, some the best at digging. This animal eats pests, this one rids the soil of impurities, that one has an amazing ability to weave grass into nests or hold its breath underwater. Dolphins, therefore, are scored as Most Intelligent.

I am probably to blame for the students' insistence on this categorization, because of course I emphasize the amazing adaptations of animals when I teach, adaptations that do indeed allow animals an incredible range of abilities. But the categories hark disturbingly back to the *scala naturae* and suggest that woe betide a species which, like an untalented child, doesn't have anything it "does" better than anybody else. Furthermore, the categories move us toward placing an economic or intellectual price on every species. If we are to preserve animals by arguing their worth in terms of drugs or food or keeping pests in check, we are forced to search for a unique trait to assign to each and every species. Similarly, must we find a role for each species before it is to be permitted to persist?

The idea that all species have a talent which is assigned to them and which thus—and only thus—renders them valuable can easily backfire in several ways. Many conservation biologists worry that what are called "charismatic megafauna" become overemphasized to the public, while simultaneously recognizing that journalists will film the release of condors into the wild more readily than the egg laying of one of the few remaining endangered Quino checkerspot butterflies *(Euphydryas editha quino)*. Our zeal to categorize animals may have made them more vulnerable rather than less. Anyone with a job can be fired. I am concerned that people will lose interest in cetaceans when they are no longer viewed as icons, and this would be unfortunate in light of increasing pressure from Japan and Norway to resume whale hunting. And what if we decide that intelligence should be defined in a way that excludes the whales, perhaps because they lack a material culture, perhaps for some other reason? Some of the most ardent proponents of cetacean conservation caution against creating this hierarchy of animals to be loved and preserved, but it is not clear that many are listening.

An interesting twist to the saga of our love for cetaceans is that we have been largely unsuccessful in learning much about their sex lives. In addition to being generally hard to observe, both male and female cetaceans keep their sex organs hidden internally except during mating, an obvious streamlining adaptation. The sexes can be difficult to distinguish in the wild. This ambiguity had a somewhat ironic result: there we were in the Summer of Love, advocating a sexual revolution while championing creatures with invisible genitals. We loved them for their minds. Dolphins are still interesting and popular, of course; recent work on their ability to recognize themselves in mirrors has rekindled some of our enthusiasm. But the grip on our emotions is absent. I sometimes wonder if dolphins are familiar and slightly dull icons, like Pooh, while bonobos are the new and trendy fashion, like Pokemon.

Enter, then, a creature whose genitals, along with its sex life, are far from invisible. I think if you had asked for a counterpart to the dolphin which had all of its cuteness and fewer drawbacks to studying and working with it, and which lived in a part of the world at least as romantic and endangered as the deep sea, you could not have done better than the bonobo.

SEXUAL EQUALITY IN THE RAINFOREST

Susan Block, mentioned earlier, calls bonobos "the horniest chimps on earth," and virtually everyone who has studied them is struck by the primary role that sexual behavior plays in their lives. Like their close relatives the chimpanzees, bonobos live in large and somewhat fluid groups containing both males and females, but bonobos tend to be less violent, and female bonobos, unlike female chimpanzees, can dominate males and chase them away from food at least some of the time. The earlier designation "pygmy chimpanzees" has been dropped, in part because the two species overlap considerably in size. The apes have been studied in captivity, perhaps most notably by primatologist Frans de Waal and his colleagues and students. De Waal suggests that in bonobo society, situations involving conflict or tension, such as the introduction of a new food or the presence of an unfamiliar individual, are generally accompanied by behavior that we consider highly sexual: both sexes stimulate their own genitals, they solicit brief genital contact with other bonobos of either sex and any age, and they engage in prolonged kissing, again with partners of either sex. "It is impossible to understand the social life of this ape without attention to its sex life," he writes; "the two are inseparable" (p. 4).

This high degree of sexuality is also seen in the wild populations of bonobos studied in Zaire, where research groups focus their attention mainly on two sites, Wamba, where the animals are provisioned with food to facilitate observation, and Lomako, where they are not. Although it is difficult to accustom wild bonobos to the presence of human observers (one early study clocked only six hours of observation in two years), studies at both these locations have shown a general similarity between bonobo behavior in the field and that of their captive counterparts. While the role of sexual behavior in bonobo society has received its share of attention from researchers, as de Waal's quotation suggests, much of the current literature on bonobos in scientific publications does not dwell on this aspect of their activities; for example, a recent paper suggested that unlike chimpanzees, bonobos live in an environment where food in the form of plant material such as fruit is more or less continuously available. This steady food supply is thought to allow females more time to interact and form social bonds, another striking aspect of their social lives. In many other monkeys and apes, females are less likely to form alliances than are males, which has led to a popular belief that female bonding is uncommon among nonhuman primates and hence not likely to have been a force in human evolution. Anthropologist Amy Parish found that unlike these other species, bonobo females cooperate with each other even when they are not genetically related, an unusual observation for vertebrate social behavior. While sex is central to bonobo life, that life and not the sex itself is the focus of most scientific work.

The nonscientific public, and the nonscientific literature, have no such compunction. In an undergraduate course on animal behavior, I mentioned the bonobos for about three minutes of a fifty-minute lecture; I did point out that they exhibited a great deal of sexual behavior, but went on to discuss primate social systems in general and the nature of coalitions and dominance in several species. I also discussed the mating system of marmosets, diminutive South American monkeys which often live in nuclear family groups of a male, female, and juvenile offspring. On the final exam, I asked whether bonobos or marmosets were more likely to exhibit a high confidence of paternity, the term used by behavioral ecologists to indicate the likelihood that a given male has actually fathered a given offspring. More students got it right than not (the marmosets have a higher confidence of paternity because less opportunity exists for copulations outside the group), but the way in which they worded their answers suggested that my short description of bonobos had captured their imaginations:

"(Bonobos) have a very open sexual relationship, where marmosets mate only with females and take care of offsprings."

"Bonobos interact with many individuals and take pleasure in fornicating with different individuals."

"Bonobos exhibit random sexual actions."

"Bonobos are promiscuous. They mate with any animal that they meet."

"Their sexual behavior is erratic, with sex being a social instrument."

"Marmosets are more sexually discrete than bonobos who mate with any and every bonobo in the group."

"Bonobos have sex for fun so questionable fathers."

For the record, I have never used the word "fornicate" in teaching (or in any other context that I can immediately recall). My students, however, were clearly taken with the bonobos in a way they had not been with most of the other animals I had discussed, and of course it is the sex that intrigued them. Susan Block blatantly uses bonobos to illustrate her philosophy of "ethical hedonism," and then throws in a few remarks about how endangered the species is in the wild and how worthy bonobos are of our efforts to save them from extinction. She too is charmed by the frequency and variety of their sexual encounters, which she describes in more colorful language than that used by de Waal and his colleagues, and refers to their "using sex to make peace." It is hard not to be left with the impression that a more Puritanical ape would be less worthy of our attention.

The reception of Block's enthusiasm by more conventional students of primate behavior, as well as by those interested in primate conservation, has been less than, well, enthusiastic. In an article titled "Sin County Almanac" from the online magazine *Grist,* Erik Ness expressed doubt about Block's credibility. "Is it," he asked, "the props? A brass bed, a corn snake named Eve, a stick of salami, a velour vulva puppet—these are not the traditional tools of conservation." The piece is subtitled: "Sex sells, but can it save the planet?"

Ness is not alone. On the Bonobo Protection Fund web site which is devoted to current material about study and conservation of the animals, an anonymous critique appears detailing the reasons for keeping bonobo sexuality separate from that of humans. Apparently the segue from anthropology and animal behavior to pornography is too blatant and disturbing. At a conference on human sexual nature, de Waal presented a paper on bonobo sex as an alternative to aggression; he noted that "some participants indulged in speculation along the lines that 'there is a bonobo in all of us,'" which is certainly in keeping with Block's ideas. He was openly skeptical that such speculation could really teach us much about the evolution of human sexuality. Others argue that merely advocating bonobo conservation while reciting the more titillating aspect of their behavior is an empty effort if one truly wants to preserve wildlife.

I admit to some sympathy with Block's point of view, and I also admit to being taken with the part of her web site containing an advertisement for Harvard professor Richard Wrangham's scholarly book on male primate behavior alongside one for *Rock 'n' Roll Babes from Outer Space*. It is true that if you want people to support a scientific cause, that cause has to be accessible to them, and as Block has remarked, "A little anthropomorphizing never hurt Flipper's drive for conservation." The association of bonobos with phone sex is not the part of all this that I find disturbing. It is the idea that we will love only sexy, amusing, or charismatic species, species easily humanized. We replaced dolphins, the Most Intelligent, with bonobos, the Most Uninhibited, but the implication is that other animals without such easy monikers will be ignored. Bonobos suit a trendy view of sexuality, but what happens when the trend is over? Anthropomorphism can backfire, even on Flipper. And the fascination with bonobos is dangerous for additional reasons.

TALKING APES

Two other aspects of bonobo behavior, besides their appeal to the prurient, have made them popular, and may be part of the reason that they eventually replaced dolphins as the animal icons of favor. The first is not mentioned in Susan Block's writings, but it contributes to the impression of bonobos as more like humans than other animals may be. This is the effort by the scientist Sue Savage-Rumbaugh, co-director of the Language Research Center associated with Georgia State University, to communicate with several primate species, including bonobos, in human language.

Such investigations are not new. In the 1960s two psychologists, Beatrice T. Gardner and Allen Gardner, attempted to teach a young chimpanzee to use sign language. Other work, mainly with gorillas and chimpanzees, tried to get the apes either to sign their thoughts or to use symbols for objects and ideas. Many of the pupils, like Koko the gorilla, became near-celebrities in their own right, and have attracted a great deal of publicity. The research has been both stimulating and controversial. Some linguists have argued that none of the demonstrations suggest that nonhuman primates truly comprehend the use of human language, and other scientists are skeptical that teaching a bonobo to distinguish between Perrier and tap water, as Savage-Rumbaugh's subject Panbanisha has done, and then seeing her freely express a preference when queried (for the designer stuff, of course) really moves us along in understanding behavior.

The details of how the bonobos use language, and whether it is qualitatively different from other forms of human-animal communication, are beyond the scope of this chapter. Regardless of whether you think Kanzi, the bonobo who has received most of Savage-Rumbaugh's attention to date, is using grammar and language structure the way his champions claim, the research program has led to some interesting discussions about the relationship between mind and language, the nature of consciousness, and, of course, the ethics of controlling the lives of animals capable of playing jokes, mourning a dead pet, and enjoying television. It is truly incredible to read about Kanzi's behavior; one cannot do so without feeling an excitement at having glimpsed the mind of another species. I wonder, however, if the emphasis on making bonobos enter our world and use its artifacts might not detract from our ability to understand their own. And again, I am leery of using the research to "score" bonobos, instead of Most Intelligent, like the dolphins, or Most Uninhibited, as Most Likely to Talk Like Us, another designation that prevents us from appreciating the animals themselves in favor of using them to prove our own superiority.

CHIMPS ARE FROM MARS, BONOBOS ARE FROM VENUS

The second part of the bonobo story, aside from their linguistic skills, is their role in serving as a model for human evolution and thus for current human social systems. Sue Savage-Rumbaugh is a woman primatologist, a fact that has not escaped the attention of at least some with interest in the controversy. An article about her research in *Ms.* magazine portrays her as a maverick who, along with other female primatologists, has brought a more holistic perspective to animals. It refers to her "near-martyrdom" in the face of criticism, a characterization I suspect Savage-Rumbaugh herself might question—science is full of disagreement, and not all of it is pretty. But the role of females and feminists in primatology has always been highly charged, and nowhere is that more true than in the application of our knowledge about bonobos and other primates to ideas about what early humans were like.

As I have mentioned in earlier chapters, the study of monkeys and apes was dominated by men well into the middle of the twentieth century, and a focus of some of this research was the search for origins of human evolution. If humans are so closely related to apes, the reasoning went, then we ought to be able to draw conclusions about what early humans did by observing our relatives. One needs to be cautious, of course, because different species of primates have been separated evolutionarily for a very long time, and their paths of social development are almost certainly different from their ancestors', but the general idea had solid appeal. Between fossils and the behavior of both these nonhuman primates and modern human hunter-gatherer societies, we might be able to piece together a reasonable picture of what our own ancestors did many hundreds of thousands of years ago.

Enter what anthropologists call the "Man the Hunter" model. Championed by several eminent anthropologists in the early 1960s, this idea suggested that what made humans unique was the hunting activity of men. Women were too busy with children and stone hearths to go out and invent tools or develop sophisticated communication systems for cooperative hunting, so all the good stuff, like bipedalism, big brains, and tools must have come from the men. Similarly, observers of primates, particularly baboons, emphasized the dominance struggles between males in the troops, and viewed female behavior as having to do only with infant-rearing. Those studying chimpanzees pointed to the violent struggles among males over meat, and speculated that warfare had its origins in similar behavior among humans.

Numerous feminist and other critiques have pointed out the faults of these ideas, and it is not my intent to review them here. Women are now receiving a majority of Ph.D.s in primatology, and both this field and paleoanthropology have been subject to many discussions of how our biases about gender influenced what we were willing to see in both the fossil record and modern animal behavior. Baboon females, it turns out, strongly influence group movements, and many primate social groups are centered around females and their relatives, while males are more nomadic. Although the issue of which sex did (or does) what is far from settled, efforts to make the study of human origins less biased are ongoing.

The place of the bonobos in all this has been curious. Little was known about them until the late 1970s and early 1980s, after the feminist revolution in anthropology, and the apparent sexual equality and open sexual behavior, particularly between females, fed a newly politically correct view of both animal behavior and models for human evolution. Instead of seeing ourselves as chimplike, maybe we could go back to the path of making love and not war. Indeed, de Waal speculates, "Had bonobos been known earlier, reconstructions of human evolution might have emphasized sexual relations, equality between males and females, and the origin of the family, instead of war, hunting, tool technology, and other masculine fortes" (p. 2). This is one possibility; the other is that had bonobos been known earlier, we would have characterized them as more violent and warlike than we do now, simply because the paradigm of the day emphasized male aggression, which the bonobos do possess. The anthropologist Craig Stanford suggests that the stark contrast researchers now emphasize between chimpanzees and bonobos might represent, at least in part, the differing social mores of the researchers themselves. Of course, although they are our two closest relatives, neither species mirrors human beings exactly. De Waal goes on to say that "the bonobo and the chimpanzee are equidistant from us. Rather than favoring parallels with one or the other ape, there is no need to choose between the two" (p. 143).

But what about choosing between men and women? Stanford says, "The behaviors at the heart of the chimpanzee-bonobo interspecific variation—sexuality, power and dominance, aggression—are those that also lie at the center of the debate about human gender issues and what molds our own behavior" (p. 407). He also muses on an extension of feminist anthropologist Sherry Ortner's contention that "men are to women as culture is to nature" in the form of "chimpanzees are to bonobos as men are to women." In other words, the sexes are still different, and still stereotyped, but now we get to pick whether we like the old male version with the war toys or the new female one with lesbian sex and food sharing.

The problem, of course, is that such stereotypes are foolish, no matter how many pop psychology books they sell. It is ironic that the social system of bonobos is embraced when feminist visions of behavior urge an awareness of our biases, not a reordering of them. As I discussed earlier, it is not enough to keep the dichotomy of men as aggressive and women as nurturing, and simply elevate "nurturing" to as high a status as warmongering. With respect to the bonobos, as well as the cetaceans, we cannot use the animals as poster children for intelligence, sexuality, or language. Apes are relevant to understanding some aspects of human biology because humans, too, are apes, but we cannot take this too far. The noted primatologist Linda Fedigan cautioned against the "baboonization" of human society, whether ancient or modern, against seeing humans as having evolved from a group structured by male dominance and sexual inequality. I believe that "bonobofication" is equally ill advised.

Eight

THE ALPHA CHICKEN

It would seem that we have come full circle when a feminist author gives advice to a male presidential candidate on how to act more like a chicken. In the fall of 1999, the media was atwitter with the news that Naomi Wolf, author of *The Beauty Myth* and *Promiscuities,* had advised Democratic hopeful Al Gore on how to dress and act. But it was not just fashion advice or image-shaping, both of which are routine to modern politicians. Apparently Wolf was counseling Vice President Gore on how to be an Alpha Male. His position as the second-in-command, the columnists fretted, had left Gore with an air of second-best as well. The solution, it was thought, was some coaching that would make him take on more of the aura of a winner, coaching that included clothing tips in addition to ways to win the female vote. Why projecting heightened confidence and aggression had anything to do with wearing more earth-toned shirts (apparently part of the advice) was never clear. Wolf herself denies having dwelled upon Alpha versus Beta in her consultations, but as is often the case, the reality of the situation was eclipsed by reaction to it.

Many aspects of this minor flap were of interest, at least in passing, but politics aside, the part I found most arresting was that everyone knew what an alpha male was, or thought they did, and everyone seemed to agree that Gore would want to be one. According to the *New York Times,* "Alpha males dominate and lead other members of the pack, while beta males are subordinate and play a helpmate role." Life is better when you are on top, the reasoning went, and for evidence one simply had to look at animals, where dominance and aggression are the rule of the day and males spend their time jockeying for position in the herd, or flock, or school. Dominance hierarchies seem to make intuitive sense to people, maybe because of the same viewpoint that makes the *scala naturae* so compelling. Taken only slightly further, alpha males and the phenomena surrounding them can be seen as responsible for male violence against women as well as warfare. Here, too, some people point to the masculine image we have from the animal kingdom and view even the human male's more extreme aggressive behavior, such as spousal abuse, as in line with what we see in other organisms and have inherited from our ancestors. Another *New York Times* writer seemed to suggest that being an alpha male, in politics as in wolf packs, is tied to having lots of sex; why else would you want to be one? The implication in his article was that Americans will forever be frustrated because we want our politicians to be dominant and aggressive without being licentious, and such a combination is as unlikely to occur among senators as among elephant seals.

Males, dominance, and animal behavior are therefore linked in many people's minds, and here I wish to explore the connection between them by examining some of the myths about them that pervade our culture. Several recent books, including Richard Wrangham and Dale Peterson's *Demonic Males* and Michael Ghiglieri's *The Dark Side of Man,* examine the roots of male violence in society by looking at its evolutionary heritage. My goal is not to debate the "naturalness" of male aggression in humans but to see how its counterpart is viewed in animals, and to suggest that, as with motherhood and monogamy, the use of animals to uphold our convictions is doomed to fail.

HOW LIKE A CHICKEN

The scientific study of dominance in animals goes back about two hundred years. Pierre Huber, a Swiss entomologist who observed bumblebees as they established their nests, noted the fights that ensued between a queen after she laid her eggs and the worker bumblebees already present. Other biologists found similarly elaborate aggressive acts among bumblebees, and were surprised by the way in which the insects seemed to settle into peaceable roles after the squabbling was over, with one or another individual deemed to have emerged victorious.

It was not, however, until a Norwegian scientist named Thorleif Schjelderup-Ebbe published his observations of flocks of domestic hens in the

1920s that the study of dominance hierarchies got seriously under way. Schjelderup-Ebbe documented the interactions between pairs of hens and among the birds in larger groups, and found that one individual would peck another, who would retreat. The peck-er would as a result get access to desirable roost sites and food items, while the peck-ee in turn would defer to the bird exhibiting the aggression. The hens could remember the identities of several individuals for a period of weeks, and once the fights had determined which bird was the winner, actual physical encounters diminished and the relationships were acknowledged, with the loser or subordinate always acceding to the winner or dominant individual. This was important stuff in the life of the birds, as Schjelderup-Ebbe cautioned: "Fights among chickens, which are usually considered to be quite harmless, are certainly not so and do not result from a momentary whim. . . . They put a lot at stake, sometimes even their lives, in order to win."

Schjelderup-Ebbe further noted that the members of an entire flock could be assigned ranks, with hen X being dominant to hen Y and both dominating hen Z. He called this situation "despotism," and said, "there exists among birds a definite order of precedence or social distinction." He was quick to qualify the notion of a strictly linear hierarchy by pointing out the existence of "triangularities," situations in which A dominated B and B dominated C, but C dominated A. The larger the flock, the more likely it is to contain such nonlinearities; ten or fewer birds and life is more clear-cut than in a larger social scene. Even among hens, social relationships are complex and not always predictable. In what was doubtless one of the earliest comments on the genetic basis of dominance behavior, he also mused, "the tendency to social structure is in the chicken's blood . . . [it] is inherited rather than learned." It is worth noting that social groups of both the bumblebees and hens consist entirely of females, although Schjelderup-Ebbe eventually studied both roosters and birds of both sexes from a variety of species. The original expression, then, is "alpha hen," not "alpha male," though this is probably not worth pointing out either to Gore or to the *New York Times.*

These findings were received with great interest by the animal psychology community, and numerous other investigators studied interactions among chickens and other group-living animals, especially birds and primates. For obvious reasons, dominance hierarchies became known as pecking orders, and the ramifications of dominance rank, also referred to as social status or social rank, were examined under a variety of circumstances in the laboratory and field. Groups of monkeys or pigeons were placed under conditions of stress, or were given only a few choice items of food, and subsequent effects on rank observed. Eventually psychologists began studying dominance hierarchies in humans, though most of them recognized that it would be difficult to assign a single rank to a person because each of us functions in many different roles; a top athlete may not attain a high score on a physics exam. One solution has been to examine people in a restricted setting, with preschool children being a favorite set of subjects.

This is all well and good, but how did looking at barnyard hens turn into something Al Gore needed to be concerned with? And is it reasonable for dominance to have become inextricably entwined with sex in many people's minds? To answer these questions, we need to take a closer look at what dominance hierarchies in nature are really like.

MYTH 1: DOMINANCE AMONG ANIMALS IS PREVALENT, CLEAR-CUT, AND ALWAYS BENEFICIAL

The triangular relationships discovered in the chickens proved to be no exception to the generality of complex and sometimes confusing findings about dominance interactions among animals. One of the first problems in trying to understand how dominance functions in everyday life was connecting the results of laboratory tests to their natural counterparts. For example, in one popular type of experiment, two rodents were allowed to enter a plastic tube at opposite ends. The duo met somewhere in the middle, and the individual that continued moving forward and made the other individual back up was deemed dominant. Other experiments deprived the animals of some resource like heat, food, or water, and then gave a limited amount to a group under observation. The one able to possess the items was dominant. The question was, how relevant are these results to what animals do in nature? Presumably two gophers simultaneously try to occupy the same burrow once in a while, but that hardly constitutes an overarching principle of animal behavior. And even if an individual can sequester a food pellet, does this mean that it can also get the best sleeping sites, or—more important—the best and most mates?

Furthermore, both in captivity and in the wild animals vary enormously in how overt dominance interactions are in their lives and how they are expressed under different circumstances. Freshwater fish called medaka will share food in an aquarium when it is abundant, but if food is restricted, a dominant individual may become territorial and keep other fish away.

Some birds, such as sparrows and their relatives the juncos, have dominance hierarchies in the large flocks that form during winter which disappear at other times of the year.

In 1946 the great animal behaviorist T. C. Schnierla wrote worriedly, "We must seriously entertain the possibility that dominance theory is an inadequate basis for the study of vertebrate social behavior." He pointed to the risk of circularity in defining dominance; a 1938 paper on black-crowned night herons *(Nycticorax nycticorax)* had used the relative height at which the bill was held as a criterion for determining the dominant member of a pair, assuming that bill height reflected food-grabbing ability of the bird as a nestling, with little evidence to support such an assumption. A heron with a high-pointing bill may be called dominant, but if dominant just means that an individual holds its bill higher and is not linked to any other function, the notion is not much use.

The literature on dominance in animals continues to be rich but fraught with controversy. Several interesting lines of research are currently under way, including investigation of the so-called winner effect, in which prior winners are likely to continue winning, even if their opponent is one that might have beaten them under other circumstances. This effect has been demonstrated in some species, including several fishes, and is surprisingly absent in others. Costs and benefits of combat have been studied in animals ranging from crayfish to macaques, with studies still continuing to use the ever-popular chicken. Research from my own laboratory on dominance interactions among hens, not of domestic poultry but their ancestors the jungle fowl, has shown that females infested with a gut parasite are less able to achieve high social status, and that while body size plays a role in getting to the top of a hierarchy, the picture for both male and female jungle fowl is more complex than simply having the biggest bird win. The primatologist Jeanne Altmann has studied baboons in Africa for over four decades, and maintains that rank is critical in determining reproduction for females, less so for males. Shirley Strum and Bruno Latour suggest that, at least for primates such as baboons, dominance interactions are so subtle and alliances so shifting that it makes more sense to view dominance not as a fixed organization which the animals enter but as a characteristic of a broader society being continually reworked by their activities.

The question of what an animal gets out of being dominant has proved particularly knotty, surprisingly so considering the assumption most people, including many biologists, made about how the "alpha male" must get the females. Similarly, the "alpha female" (a term rare in popular use except in reference to the mate of the alpha male, even though dominance hierarchies among females are, as we have seen, quite common) was assumed to have more and healthier offspring than those lower in rank. Although in some species the male best at brute-force combat can indeed control access to large groups of females, this is beginning to seem like an exception and not the rule. The literature on male dominance in primates such as baboons has been especially controversial in this regard. Males that appear to be dominant do not necessarily father more offspring, as DNA fingerprinting has revealed. Barbara Smuts's groundbreaking work on social relationships between certain males and females in baboon troops, which she herself terms friendships, suggested that males that support females in a variety of contexts may benefit in many ways, including obtaining future mating opportunities. The best we can do at this point is summarized in this statement from a book about animal conflict: "In some primates, in some circumstances, dominant animals derive some benefit from their high status." Again, if a generalization could be made, it would apply to female monkeys, not males; female rank is fixed in many species and therefore females play for keeps; males are more difficult to generalize about. Hardly a ringing endorsement of the unlimited power of male brutality, but also hardly likely to make the evening news.

The consequences of being dominant also vary depending on whether the hierarchy is stable or marked by frequent turnovers in status. Indeed, early animal behaviorists, including the ethologist Konrad Lorenz, postulated that the function of dominance is to promote social stability within a group, making it easier for everyone to go about their business free of continually having to prove their rank to others. This idea is not unreasonable, in that both winners and losers can benefit from the absence of conflict, but it led to some erroneous conclusions about the ways in which animals refrain from fighting, as I discuss below. Nevertheless, Robert Sapolsky, who spent many years examining the physiology of dominance in savanna baboons, discovered that although males near the top and bottom of the hierarchy had characteristic hormone profiles, the relationship between the testosterone levels of the ranks depended on social stability. When the hierarchy was stable and clear-cut, testosterone levels were similar in the two types of males, but the dominant individuals did not experience much stress-induced change in hormone levels. If, however, ranks of males were perpetually changing and the hierarchy shifting, testosterone levels were much lower in the subordinate baboons. In other words, it's not just what rank you hold at the moment, it's how sure you can be of maintaining your status in the days to come.

What about the dominance relationships between the sexes, crucial to many people's models of male violence and aggression? As you might expect, among animals these too show huge amounts of variation. Males by no means uniformly dominate females, though they tend to do so in species with large disparities in size between the sexes. Among most primates, males outweigh females, and often, though not always, can chase them away from a desired resource like a food item. Some notable exceptions occur, such as in vervet monkeys, Old World primates that inhabit a wide range of habitats in Africa and have been studied by numerous scientists. Vervets live in fairly large groups and do not show a pronounced sexual dimorphism in size. Females frequently make males back down, and pairs or larger groups of females are even more successful in dominating males.

Even when males can chase females from a food dish, this does not necessarily mean that they are able to get their own way all the time. Most if not all female primates, birds, and insects will, at least sometimes, refuse to mate with males that solicit copulations from them. It is the females themselves who do the solicitation for sex in other species, such as owl monkeys. Dominance relations also change with the season, so that females may become dominant over males during the breeding season but not before or after, or vice versa. Schjelderup-Ebbe was perplexed by situations in which female birds dominated their male counterparts, because, as he put it, "female despotism constantly brings about degeneration through the hindering of pairing, thus operating *against* the increase of the species." In other words, he was concerned that the reluctance of females to mate might result in their not reproducing at all, a problem which is without basis in reality, as we see in other chapters. One gathers that Schjelderup-Ebbe was not a proponent of the active female role in mating, though he did offer the consoling thought, "The male is not cruel as a pairing despot." The nature of mating itself led many researchers to link what they observed—male advances followed by female retreats—to a power imbalance, with males assumed to be dominant in sex if they began a sexual interaction. I suspect this interpretation has more to do with the biases of the observer than the reality of animal mating; for one thing, females frequently initiate sex, and for another, activities in courtship cannot always be generalized to activities in other arenas, as I discuss below with respect to predation.

The myth, then, that dominance is so widespread, and so uniformly male-oriented that it renders hopes for sexual egalitarianism in our own species worthless, is just that, a myth. It is true that male and female interests differ, but this is far from saying that male interests will always predominate. As for Al Gore, being alpha may be less easily defined than journalists would like to think. A faint consolation, given his eventual loss, is that at least none of them started calling him hen-pecked.

MYTH 2: FEMALES ARE RARELY VIOLENT OR AGGRESSIVE

I discussed this topic in the chapter on ecofeminism, but I return to it here because I want to emphasize some of the more subtle ways in which females compete both with each other and with males. Females do compete directly, and rank is critical in the lives of many female primates as well as the wasps, bluebirds, and reed warblers mentioned earlier. The discovery that many monkey societies are matrilineal, with the rank of the mother being passed on, more or less, to daughters, was one of the major advances in primatology.

Although the details differ both among species and between the sexes, broadly speaking females compete and are aggressive over the same things that males compete and are aggressive over: resources that make it more likely an individual's genes will be passed on into the next generation. Sometimes this competition means that a given female will do better if other females around her cannot reproduce. In a variety of animals, dominant females exert such control over the lives of subordinates that the subordinates become physically incapable of producing offspring. The classic case is that of the social insects, in which the queen lays the eggs of all future workers (except for some intriguing cases of worker insurrection, which I cannot discuss here), but the situation among mammals brings this situation a little closer to home.

Dwarf mongooses are small carnivores that look a little like weasels (although they are not in the same family). They live in arid regions of Africa in family groups containing from three to eighteen individuals, but only one female produces most of the young. The nonbreeders assist in various caretaking activities, such as defending the colony from predators and digging in the burrow system used by the group. They also help rear the young, but being mammals, baby mongooses need milk, and this is usually a product only the mother can supply. Occasionally, however, a subordinate female becomes "pseudo-pregnant," and undergoes hormonal changes as if she were actually carrying a fetus and later were to give birth. This results in her producing milk, which she uses to nurse the young of the dominant female. These "superhelpers" are virtually always full siblings of the young they nurse, which means they are helping to rear individuals sharing some of their genes, a partial compensation for their sacrifice. In addition to them, some females do reproduce along with the dominant breeder, though only 13 percent of subordinate females produce litters. When they do, the young are reared communally along with the young of the dominant female. The dominant female seems to allow those subordinates to breed that are either the greatest threat to her own status (because they are older and larger, and more likely to be physically dominant) or most likely to leave (less closely related and so less likely to gain by staying and helping). This small share in the reproduction is thus almost a bribe given to certain females by the dominant one. DNA fingerprinting has shown that reproductive sharing also occurs among the males, with subordinates siring 24 percent of the offspring, mostly by copulating with the dominant female who then produces a litter of mixed paternity.

Similar reproductive suppression in subordinates has been documented in tamarins and marmosets, tiny monkeys that live in the treetops of South American forests. The mechanisms by which the dominant females exert their influence have been intensively studied by the behavioral endocrinologist Jeff French at the University of Omaha, and at least in some species it appears that odors produced by the dominant female help suppress ovulation in subordinates; if subordinate females are removed from a social group, but still exposed to urine-soaked material from the dominant female, they continue to show atypical and apparently infertile menstrual cycles. Conversely, normal ovarian function may appear in the subordinates if they are left in a group but the dominant female is removed. Reproductive suppression is a classic example of the mind-body link, or at least of the effect of behavior on physiology.

Like other characteristics of dominance, reproductive suppression has costs and benefits for all concerned. Obviously, not reproducing is detrimental to the subordinate females, but often they have no choice, because the environment is such that going off to reproduce on one's own presents too many risks to be undertaken. In some cases, subordinates do indeed breed, and the dominant individual must either tolerate the mutiny or drive the subordinates off and face losing their help in the social group. As with other forms of dominance interactions, these relationships between group members are complex and may be continually reevaluated.

The point here is that despite the absence of overt violence, female dominance can have far more ruthless consequences than the most savage canine-baring fight between male baboons. Being prevented from breeding is the most serious consequence of any behavior, and it is foolish to conclude that because females do not have physical fights as conspicuously as males (and they do, in fact, have such fights at least sometimes) their competition is less keen. Again, however, this does not mean that male dominance in human society is necessarily an easily altered artifact of culture, or that women are just as brutal as men. It is nevertheless wrong to argue that men are "naturally violent" and that we should not be surprised when they batter women because we find the roots of violence in male animals and not females. The roots of violence are not found in males alone.

MYTH 3: HUMANS ARE THE ONLY ANIMALS THAT ROUTINELY KILL THEIR OWN; OTHER ANIMALS REFRAIN FOR THE GOOD OF THE SPECIES

Konrad Lorenz, the Nobel Prize–winning ethologist, observed animal conflict in a variety of species, including dogs and wolves. He noticed, as had many others, that after the fight has continued for some time the weaker individual "holds away his head, offering unprotected to his enemy the bend of his neck, the most vulnerable part of his whole body! Less than an inch from the tensed neck-muscles, where the jugular vein lies immediately beneath the skin, gleam the fangs of his antagonist from beneath the wickedly retracted lips" (*King Solomon's Ring,* p. 186). Although it would seem that this is the moment of death for the throat-baring wolf, in reality an extraordinary thing often happens. The two opponents remain in this position for some minutes, and then the winner simply turns and walks away, leaving his enemy unharmed. Lorenz found this restraint impressive, and suggested that humans would do well to emulate it. He feared that the advent of manufactured weapons such as guns and bombs removed opponents from the lupine form of direct confrontation and therefore circumvented the kind of social inhibition he found so laudable in animals, making it easy for people to kill each other carelessly and without regard.

While it is certainly true that modern weaponry enables wholesale slaughter on a scale hardly possible in the average wolf pack, Lorenz went further in his interpretation. He suggested that the wolf is refraining from slitting his opponent's throat because such behavior, while it would benefit the winner, is not good for the species. Widespread carnage is simply too destructive to the population to endure, and therefore animals have evolved a way around it.

This argument seemed logical to many of Lorenz's contemporaries, and it always seems logical to my animal behavior students as well. After all, why can't we, as the saying goes, just get along? Wouldn't everyone be better off if no one was a bully? This viewpoint also suggests that animals have it right and humans have it wrong, unlike beliefs about the nobler impulses of civilized humanity that allow us to rise above the brutish dominance relationships of animals. Proponents of it sometimes go further and gravely intone that man (and I use the word advisedly) is the only animal that regularly kills his own kind.

Let us examine this notion, as to both the truth of the last statement and the logic of the Lorenzian idea of social inhibition and restraint from violence. Imagine a situation in which the wolves fought as described, but the winner savagely attacked his submissive partner and, indeed, sometimes killed him after his throat was bared. The winner then went on to enjoy the fruits of his victory, whatever those might be. Presumably, if dominance is beneficial, this would mean that at some level he attained higher reproductive success than other, more subordinate, wolves. Now compare that outcome to one in which the winner follows Lorenz's rules about social inhibition for the good of the species. Both winner and loser live to fight another day, which means that the winner risks losing his dominance and the advantages that accrue from it. A little thought reveals that ruthlessness will pay off more than restraint: a killer wolf will leave more offspring, who are also likely to be killers, than a merciful wolf, all else being equal, simply because the merciful wolf might get killed himself later on. Some generations later, the killer characteristic will prevail and the merciful individuals, along with their genes, will have finished last in regrettable good-guy fashion. Mercy may be good for the rest of the wolves, but if it doesn't benefit the individual performing the behavior, natural selection cannot act on it. This realization of the level at which selection generally occurs is one of the great insights of modern evolutionary biology.

So why, then, did Lorenz observe what he did? Why does the wolf stop before killing his victim? The key lies in the *ceteris paribus* clause. All things are not equal, because escalating a fight to the death has potentially serious consequences for all concerned, not just the loser. Combatants have the means to injure each other in all animal species in which physical conflict occurs, and even the winner can emerge from a battle with life-threatening injuries that make victory hollow. The wolf that walks away from a more intensified fight spares his opponent, to be sure, but also does not risk the cost of such a battle to himself. Social inhibition can be just as selfish as blowing one's opponent away. Everything depends on the costs and benefits of progressing with the struggle.

Biologists have examined how decisions about escalating versus abandoning a fight should be made; one of the most useful tools for the research has been a body of mathematics called game theory, in which the various strategies are assigned points representing fitness and the results of interactions calculated under different circumstances. One of the classic "games" is called Hawk-Dove, which has players called Hawks who always escalate when challenged, as opposed to the Doves, who always back down. Whether being a Hawk or a Dove is better depends entirely on the payoffs associated with each type and on the proportions of each type in a population; being the only Hawk in a population of Doves leads to great success for the Hawk, but conversely, being the only Dove in a population of Hawks gives more points to the Dove because each confrontation between two Hawks is likely to be costly to both of them. The games can become quite complicated, with one version containing a third strategic position, called Bourgeois, who sometimes plays the role of Hawk and sometimes that of Dove.

What this means is that sometimes being aggressive pays off and sometimes it doesn't, an unsatisfying but accurate statement like the one about dominance in primates that I quoted above. Carnage is likely to occur in animals, but only when the stakes are high and winners can win a great deal. The perfect example of such devastation occurs not in wolves or lions or even real, rather than theoretical, hawks, but in a creature so tiny it can be studied only through a microscope: the fig wasp.

Over six hundred different species of fig wasps exist, and all are essential to the life cycle of one or more of the numerous types of fig trees that occur throughout the world. Part of the wasps' natural history includes a period inside the maturing fruit, where male wasps vie for an opportunity to fertilize the females. One form of the males is wingless and cannot leave the fig, and therefore their only opportunity to pass genes into the next generation is to monopolize the females they encounter. They do so with remarkable violence; specialized jaws allow decapitation of a rival with a single lethal snap. The evolutionary biologist W. D. Hamilton documented the slaughter inside one fig and found fifteen females, twelve uninjured males, and forty-two other males either dead or dying from wounds leaving them limbless, headless, or otherwise torn to shreds from the battle. (In case you were wondering, yes, this means that some commercial ripe figs contain real wasps. Others are artificially pollinated. But don't worry; you've been eating them for years, and they taste exactly like figs.)

So much for our gentler animal cousins that kill only when necessary for food. Similar butchery can be seen in other species, vertebrate and invertebrate, male and female. Richard Wrangham and Dale Peterson argue that only in humans and chimps do killers seek out and "deliberately" destroy their victims, but while this may be the case among primates, it is hardly unique in the animal kingdom. It is perhaps more likely that males will fight harder because the stakes, in competitions for females to fertilize, are often higher for them; as I discussed, females are limited in reproduction not by the number of males they can find to inseminate them but by the number of offspring they can produce and help to survive. But it is the stakes, and not the sex, that determines the rules.

MYTH 4: MALE AGGRESSION IS TIED TO HUNTING PREY

Some anthropologists and psychologists have traced human aggression and male violence to an early history of hunting, and at least one social critic, Barbara Ehrenreich, to a history of being hunted. Locating a living animal, stalking it, and killing it are thought to represent aggressive acts, and because in the prototypical human hunter-gatherer society men are supposed to have done much of the group hunting, the reasoning goes that natural selection for good hunters gave us, willy-nilly, hostile men. Some people have taken a slightly different tack and suggested that the acts of courtship and sex itself, whether in humans or other animals, stem from behaviors essential to hunting, in which (can you guess?) the man is a predator and the woman his prey.

One of the earliest of these proponents was Havelock Ellis, the psychologist who wrote the landmark *Studies in the Psychology of Sex* and was active in the early part of the twentieth century in the United States. Ellis's main thesis, that sexuality represents a power relationship with males dominant and females submissive, has been roundly criticized by feminists and other scholars, and his ideas, while undeniably influential, are not so widely held today. I am more intrigued with his use of animal behavior to inform his views. He thought that in animals, the act of copulation represented the hunter grasping his prey, which was why males were on top of females from behind; in humans, however, "The . . . male may be said to retain the same attitude, but the female has turned round; she has faced her partner and approached him, and so symbolises her deliberate consent to the act of union." Whether in animals or humans, Ellis thought, a certain female reluctance was to be expected, and the male in turn expected to pursue just as he pursued wild game. Similar parallels between courtship and predation are drawn by several of the early ethologists, who also noted that some displays exhibited by some mammals and birds before mating are very much like postures that are part of hunting food. Anthropologist Melvin Konner makes a comparable suggestion in his book *The Tangled Wing*, suggesting that this relationship is what makes women initially respond to overtures of sex with fear.

I wonder what Ellis would have made of field crickets, in which females, the larger sex, mount males to receive spermatophores, small packets of sperm, which must be painstakingly threaded into the female's genital aperture and cannot be forcibly given. Other species of crickets must align themselves tail to tail, with rear ends facing each other. Presumably he never saw insects mate, nor perhaps many animals other than domesticated mammals. In reality, though frontal copulation is not the rule, it occurs in a variety of animals, and while it may appeal to us that bonobos, orangutans, and some cetaceans engage in it, in itself it is not particularly significant. Courtship may in fact superficially resemble predation, because one party often advances while the other party retreats. I suppose one could say that because a conflict of interest occurs between the sexes, the similarity is even greater, since obviously the hunted and hunter disagree about the best fate for the prey. But it is false to conclude that therefore females say no when they mean yes and males have to exert their will for it all to work out appropriately.

The first problem is that, as I have already discussed, females may be evaluating males during courtship to ensure that their mate is a good one. This may necessitate watching as he performs elaborate courtship dances and other behaviors; the male is expected to try and cut these short and attempt to mate, because he has nothing to lose and everything to gain by doing so, but the female is equally expected to make him wait until her information is complete, perhaps by retreating, perhaps by refusing to allow him to shortchange her. This is conflict, but it is not a miniature version of predation. It is no more indicative of the pursuer being dominant to the pursuee than a job candidate strutting his or her abilities before the boss should be.

The second problem is that hunting is a more widespread and less glamorous profession than it is sometimes made out to be. We tend to think of predators as animals that subdue relatively large, usually warm-blooded, prey, but there is no a priori reason to dismiss insectivores like, say, warblers or hedgehogs, from their ranks. Some biologists refer to any discrete food item as "prey," and talk about animals such as the seed-eating kangaroo rats as seed predators. Even if that is going a bit too far for some, is it any less savage to bite a worm than a weasel? Why is a hawk swooping down on a mouse seen as more aggressive than a songbird snapping its bill against the hard shell of a beetle? Hunting is getting food; it is not waging war. To be sure, group hunting such as that seen in chimpanzees and many human societies does involve elaborate behavioral rituals, and in some cultures hunting, because it requires bravery when the prey is itself dangerous, is used as a test of manhood. But this does not mean that predation itself is aggressive in all its forms, regardless of whether it resembles the relationship between the sexes.

Finally, even if courtship were derived from predation, and even if predation were aggressive, both ideas I just cast doubt upon, the fact remains that in virtually all animals, males and females both hunt. In lions, of course, females do most of the hunting; male violence is still present, but it is directed toward the rival males and their offspring, and infanticide is common. The role of male hunting in human evolution is the subject of hot debate among anthropologists. But in animals that do not hunt cooperatively, males simply do not go out and bring home the bacon, or the seal meat, or the caterpillars, while the females stay home with the kittens, pups, or chicks. Any tendencies toward aggressive behavior that came from hunting food would have to occur in both sexes.

It is undeniable that aggression, violence, dominance and war are all, to use a trendy word, gendered in our society. That is, they all have connotations with maleness and femaleness, and I am not suggesting that human aggression is just as common, or is viewed the same way, in women as it is in men. I wrote part of this chapter during the Super Bowl, and no other activity could make clearer the glorification of male violence. Males and females have, as I have emphasized repeatedly, been subject to different kinds of selection both in our own and other species, and male competition is often valued. But aggression means much more than a mindless striving to be alpha, something that should be as clear to Al Gore and the rest of us as it is to the average chicken.

PART THREE

Human Evolutionary Perspectives

Nine

SOCCER, ADAPTATION, AND ORGASMS

IF OUR HUMAN BODIES EVOLVED, with their opposable thumb and atavistic appendix, what about our behavior, especially our sex roles? As I discussed at the outset, the sociobiology controversy raged, not about the selfish genes of clams or carnations, but about our own genetic tendencies to love or war. I have been arguing throughout this book against the simplistic use of animals as role models and also against the assumption that we are like the animals we see. But what about humans? Are some of the same dangers and biases apparent in examining ourselves? In the chapters that follow, I explore a few of the more controversial aspects of human behavior that have to do with sex and gender, and suggest that some of the same difficulties we have with seeing what animals do plague the understanding of our own behavior. In some cases, not only have we been drawing some flawed conclusions, we have not even been asking the right questions.

In his famous complaint, Henry Higgins asks (rhetorically, one assumes), "Why can't a woman be more like a man?" Questions like this one are rarely asked simply to gain information, and indeed *My Fair Lady* illustrates a pervasive impatience in society with the functioning of the female sex. Nowhere is this frustration more apparent than in discussions of female sexuality itself, which have run the gamut over the years from the wandering hysterical womb to the G spot. Here too nonhuman animals have played a part, and in this chapter I examine the beleaguered role of the female orgasm and what it means for our views of the evolution of female sexuality.

Most biologists and psychologists agree that female and male orgasms are fundamentally different in at least some respects. This difference seems to stem from the plumbing, both anatomical and physiological: males have a penis, which is where both sperm and orgasm arise. Orgasm is generally inextricably entwined with ejaculation, and fertilization for men is therefore tied to orgasm as well. Not so, of course, for females, who can conceive without orgasm. Sexual pleasure is centered on the clitoris, which may or may not be stimulated during intercourse. The clitoris and penis derive during embryonic development from the same tissue, which is then modified to produce the sex organs of a boy or girl. This is why ultrasounds during pregnancy cannot reveal the sex of the fetus until it arrives at a particular stage of development; it's not that the penis is too small to see, it actually has not yet appeared.

These simple facts took a tortuous historical path to discovery and are still the cause of a fair amount of mental and physical grief. Some of the blame for the former can be laid at the feet of Sigmund Freud, who declared that women's orgasms that centered on the clitoris were infantile and represented an immature stage of development; only orgasms achieved via penis-in-vagina intercourse, with no additional messing around in private parts by other appendages, were considered worthy of a healthy woman. This famous pronouncement led to feelings of inferiority or pathology in many women whom we would now count as perfectly normal in their sexual responses. It also led to a virtual epidemic of "frigidity," the inability to respond appropriately during sex, a disorder that was as relevant to the political position of women as it was to the one they assumed in bed. Freud was oddly ignorant of biology in a number of ways besides this one, and the social significance of his stance on this and other issues relevant to women has been discussed thoroughly elsewhere. What is interesting here is that his common assumption about female vs. male sexuality has persisted, not only in ideas about the psychology of gender, but in ideas about evolution and sexual behavior in humans and nonhumans alike.

As I discussed in an earlier chapter, people often assume that the male version of things is the norm, and females are a variation, different, a fillip added on afterward. In the case of orgasms, this means that the male way is the Way It Is Supposed to Be. From there it is a small leap to the conclusion that if females differ from males in their sexual response, they have a problem. This problem seems to be that if women's bodies operated as they should, so-called straight intercourse would always result in female orgasm along with that of the man. Even after Masters and Johnson supposedly debunked the separation of clitoral and vaginal orgasms, the scientific literature as well as popular media was full of advice and hand-wringing about the best way to help women do what seemed to be so natural and easy for men. Their work, published in the mid-1960s, was a landmark, and represented an enormous advance over the notion that female sexual pleasure was nonexistent, or at least unimportant. Now, however, the view that women may be handicapped, but deserve to be helped, still seems to carry with it the notion of an ideal, a model system for sexuality itself. The situation was complicated by a lack of information, still present today, on what constitutes a common, let alone normal, sexual response in women. Although the four-stage Masters and Johnson categorization of the sexual cycle was embraced with enthusiasm, many researchers have since pointed out that, by necessity, the sample of volunteers with which they worked was not exactly a random representation of humanity. Relatively few people are willing to have their every gasp and secretion monitored by machines in a laboratory. The conclusion that men and women have the same sexual response pattern is perhaps premature, and therefore the suggestion that women can and should be helped to be more like men is suspect as well.

It is undeniably true that the clitoris is located where it may not be stimulated to orgasm during intercourse. The significance of this fact has been treated differently by different people, however. Freud took anatomy and made it into philosophy, and misogynistic philosophy at that. Other psychologists have had a variety of interpretations. What about evolutionary biologists?

ARTIFACT OR ADAPTATION?

The evolutionary significance of human female orgasm has received quite a bit of attention over the last decade and a half. Several scientists mused on ways in which female orgasm might help those women experiencing it to achieve higher reproductive success, either currently or in our evolutionary history. But two considerations are important here. First, before examining the adaptive significance of a trait, we need to determine whether the trait is an adaptation at all. Take, for example, the nose. In a classic paper on the pitfalls of wanton application of the theory of natural selection to many traits, the evolutionary biologist Stephen Jay Gould and the geneticist Richard Lewontin pointed out that although human noses are shaped in a perfect way to support spectacles, no one would suggest that the bridge of the nose evolved because of selection for that function. The nose is not an adaptation for holding up glasses, and it would be absurd to look at different people's noses to see which of them represents the organ best suited for the task and conclude that such individuals have been shaped, so to speak, by natural selection. To qualify as an adaptation, a characteristic must have been selected for a particular use. Because we cannot go back in time and view the ancestral form of most traits, it is often difficult to know whether any particular trait is an adaptation or whether, like the bridge of the nose, it was co-opted for a different use much later in its history.

With respect to female orgasm, the question is whether there is anything to explain at all. Gould weighed in with the suggestion that female orgasm is a by-product of the clitoris developing from the same embryonic tissue as the penis. Hence, he argues, females have orgasms because males do. He criticizes Freud for attaching values to the supposedly more mature vaginal orgasm, but sees nothing about female orgasm per se that makes it adaptive. Therefore the "problematic" position of the clitoris isn't a problem at all. Female orgasms are not adaptations to begin with; they are carried along through developmental inertia because selection has not removed the clitoris from females.

He also chides Sarah Blaffer Hrdy, primatologist and researcher on biology and gender, for her claim that orgasm evolved when our primate ancestors had more than one sexual partner during periods of sexual receptivity. Hrdy, in her book *The Woman That Never Evolved*, suggested that females seeking multiple mates to satisfy their sexual needs would also achieve the goal of confusing paternity; males could not determine if they were the father of a given female's offspring. Such confusion can be adaptive because in at least some mammal species, males commonly kill the offspring of females they have not mated with, a behavior that brings the female into estrus more quickly and therefore enhances the male's own reproductive success. A female driven to seek many copulations in rapid succession because of the proximate reward of sexual pleasure could therefore have more surviving offspring because they would be safer from infanticidal males. Hrdy thus deals not only with female orgasm itself but with the capacity for multiple orgasms without lengthy refractory periods in between. In her view, female orgasm is indeed an adaptation, at least under circumstances where females benefit from mating with multiple males. Gould's problem with this scenario is that it seems driven to seek a function for a trait that may not have one; why bother to construct what he has termed just-so stories to rationalize the existence of traits that are merely artifacts or remnants of developmental constraints?

Sarah Hrdy also notes that an examination of the clitoris among primates reveals it to be extraordinarily variable, just as the penis is, but in its own independent way. Among the great apes, for example, chimpanzee penises are long and thin, while females have very large clitorises. The bonobo penis is similar to that of the chimp, but the clitoris is crescent-shaped, and frontally placed, perhaps because selection favored a position maximizing stimulation during the genital-genital rubbing common among females. Finally, the human penis is quite thick, but the clitoris is smaller than in chimps. This does not seem like a case of the clitoris demurely following where the penis has led, evolutionarily speaking.

Orgasms are hard to measure objectively, and clearly vary in humans depending on environmental circumstances. It is obviously impossible to hunt back in time for the primordial climax, as is the case for most traits that do not preserve and fossilize. How else might we establish whether female orgasms (or other behaviors) are an adaptation? One suggestion has been to examine the behavior in question and see whether it exhibits specific components that could not have arisen unless selection had acted upon it. It is easy to see that nothing about a human nose can only be explained by selection on its structure to hold up glasses; other functions can more parsimoniously (and logically) explain the same structures. John Alcock, a behavioral biologist and writer of popular books as well as textbooks, responded to Gould by stating, "Female orgasm is *not* an imperfect, half-hearted imitation of male orgasm, but a strong physiological response that is different in pattern and timing from male orgasm." In other words, he thinks female orgasms look as though they are selected to operate in a certain way, in females, and do not show indications of being an artifact of selection on males. He turns the apparent difficulty of reliably achieving orgasm in heterosexual intercourse into a sort of discriminating palate argument; women are testing the consideration of their lovers and will remain with those who care enough about them to ensure their sexual pleasure. Those men in turn are likely to be good partners and involved parents, presumably along the lines of, "If he makes sure I enjoy sex and have an orgasm he'll be more likely to take the kids to Little League."

Gould therefore does not think female orgasms are an adaptation, while both Hrdy and Alcock do. An additional complication arises from an important distinction between the selective forces responsible for the origin of a trait and those causing the trait to persist. For example, the feathers on birds are generally agreed to have arisen as thermoregulatory devices; they helped keep bird ancestors warm by providing insulation. They therefore almost certainly occurred in birds before flight itself evolved, and certainly before birds like peacocks spread their tail feathers to display brilliant colors attractive to females. Is it currently adaptive to have feathers, from the perspective of flight and mate selection? Of course. But the trait's origin is a different story. So even if a trait is currently adaptive, it may not, strictly speaking, be an adaptation, because it evolved owing to selection for a different function. With respect to orgasm, Gould and Alcock are at cross-purposes, because Alcock is proposing a way in which it can be construed as adaptive, regardless of its origin, and Gould feels that it is sufficient to conclude that it originated as a by-product of selection on males.

Other ideas about female orgasm have also generally fallen into either the adaptation camp or the artifact camp. One of the earliest writers about the evolution of human sexuality is Donald Symons, an anthropologist at the University of California at Santa Barbara. He pointed out that two schools of thought have characterized ideas about the evolution of female orgasm. First is the idea that it is unique to humans, and is either important in maintaining the pair bond between a man and a woman or otherwise socially useful. Second is the suggestion that female mammals generally experience orgasm, and females have a capacity for sexual enjoyment for its own sake. Neither of these ideas seems likely to him; he states, "If . . . adaptive design can be recognized in such features as precision, economy, and efficiency, it seems to me that available evidence is, by a wide margin, insufficient to warrant the conclusion that female orgasm is an adaptation" (p. 89). He agrees with Gould and suggests, "The female orgasm may be a byproduct of mammalian bisexual potential; orgasm may be possible for female mammals because it is adaptive for males." A bonus, nothing more. Furthermore, "The ability of females to experience multiple orgasms may be an incidental effect of their inability to ejaculate" (p. 92).

This seems to me to be a clear case of efficiency being in the eye of the beholder. Why is it less "efficient" for women to have orgasms before, after, and not always during intercourse? Why do they have to have one (or more) every single time a man does? And most of all, why is it such a problem for penis-in-vagina intercourse not to reliably result in orgasm for all women, all the time? As the psychologist Carole Wade, quoted in Tavris's *The Mismeasure of Woman,* points out (this is where the title of this chapter arose, for those who were wondering), "Sex is not a soccer game. The use of hands is permitted." This situation does not appear to be a devastating defect for many women or their partners, but it has distressed numerous biologists. It is as though they were concerned mainly with sex for amputees.

A problem with viewing female orgasm as a trait that arose only through selection on males is the same one I discussed earlier, that our biases sometimes cause us to assume that males are normal, and females are variants. What is so perfectly efficient about male orgasm, after all? Men, generally speaking, do not ejaculate the instant their penis is inserted; in virtually all mammals a period of multiple intromissions or thrusting is required. Why is that not inefficient? Because that is the way sexual behavior "is." Female sexuality is maladaptive only if the male experience is viewed as the way it is all supposed to work. Sarah Hrdy, reacting to a statement of Symons's, suggested that "The notion that woman's orgasm is 'in an evolutionary sense a "pseudo-male" response' appears to be a vestige of Victorian thought on the subject." Certainly the idea that not only do women enjoy sex but it is natural for them to do so has been historically controversial, to say the least. Many researchers, Hrdy among them, point out that female experience of orgasm varies considerably across cultures. Some groups have no apparent concept of orgasm for women, while in others, like the Mangaians, a much cited South Pacific society, women are expected to have orgasms with each sexual encounter and men supposedly know more about female sexual anatomy than most European physicians (for unknown reasons this last is virtually always the way it is phrased; I do not know who originated this wording). Or, as sex writer Susie Bright, aka Susie Sexpert, puts it: "As it turned out, the number one sexual concern of most women is that they don't orgasm, or that they can't control when or how they do. Men do not lament that they don't know how to orgasm, or if they have ever gotten off. Never. That's not biology; that's oppression" (p. 14).

Bright (or Sexpert) therefore falls on the side of those who wonder why we are making such a big deal out of women having orgasms only via a restricted method (intercourse without supplemental stimulation of the clitoris). She too sees no need to make sex like soccer, and figures that once we know something about female sexual response, the so-called problem disappears. Indeed, one of the reasons the society in which males are knowledgeable about female sexual anatomy is so striking is that in many cultures such knowledge is rare even among women themselves. This ignorance starts quite early in life in most Western societies, in which even enlightened parents, proud of their uninhibited sex education, are likely to teach their little girl that Joey has a penis and she has a vagina as its counterpart. They are unlikely to teach her that she has a clitoris rather than a penis. Symons finds the cultural variation in experience of orgasm emblematic of a by-product, an artifact rather than an adaptation; if it enhanced reproductive success we would all do it, in all cultures and all circumstances. The discussion of adaptation in humans, however, always comes with a social context. If we keep assuming that females, including variations among them, are not the norm, it will be hard to conclude that their responses are adaptations.

JACKPOTS, THE POLEAX, AND OTHER FUNCTIONS OF ORGASM

If we assume that female orgasm is adaptive (whether it originated or later became that way), several different explanations still exist for its possible utility in evolution. Turning an apparent deficit into an asset, the psychologist Glen Wilson, cited in Lynn Margulis and Dorion Sagan's book *Mystery Dance,* proposed that women should be expected to seek orgasm more assiduously precisely because it occurs unreliably. This "jackpot theory" rests on the well-established psychological finding of the rewarding nature of intermittent reinforcement; rats will press a lever more persistently if they are rewarded only occasionally than if they receive food pellets on a regular basis. Intermittent reinforcement has been used by some psychologists to explain the human fascination with gambling. Be that as it may, the jackpot theory does not explain why males and females should differ, much less why females should require, in an evolutionary sense, more sexual motivation than men.

In their book *Human Sperm Competition,* Robin Baker and Mark Bellis put forth two competing hypotheses which they purport to test with available evidence. They ignore what they refer to as noncopulatory orgasms, that is, those occurring before, after, or without intercourse, because they are primarily concerned with the evolutionary effects of sex on conception.

The two hypotheses are rather inelegantly called the poleax and the upsuck, with the latter at least giving a mental image of its meaning.

The poleax hypothesis has been suggested by several authors, including Richard Duncan, though the name appears to be original with Baker and Bellis. The idea is that once humans evolved to become bipedal, the position of the vagina relative to the ground meant that semen could readily flow out of the female reproductive tract if the woman leapt to her feet immediately after the man ejaculated. Hence orgasm serves to keep women lying down, presumably in a muzzy state of post-coital bliss, and allows the sperm to swim toward the ovum unimpeded by gravity. It does not take a lot of thought to debunk this one, at least as a governing force selecting for the evolution of orgasm. For one thing, it ignores the potential existence of female orgasm in nonhumans. For another, it predicts a strong correlation between orgasm and conception, when the lack of such a connection is what led to much of the speculation about the evolution of orgasm to begin with. Finally, although many researchers and physicians have suggested that fertility can be enhanced either by having the woman lie with her knees bent and pelvis tilted after sex, or by assuming certain sexual positions, little evidence exists to support such advice.

Baker and Bellis are more enthusiastic about upsuck. As the name suggests, the hypothesis proposes that the contractions of the vagina and uterus during orgasm help semen move into the reproductive tract, and therefore can serve as a kind of sperm aid for certain males. The authors also suggest that women may be able to selectively retain or eject the ejaculates of different men. This idea is not as far-fetched as it might appear; in many bird species, virtually the entire ejaculate may sometimes be emitted from the female's cloaca shortly after copulation, via mechanisms that are not well understood. Baker and Bellis speculate, sometimes to a degree criticized sharply by anthropologists and evolutionary biologists, about the circumstances under which this type of encouragement might be expected to occur. Indeed, whether orgasm helps or hinders fertilization is still unclear. A few physicians have proposed that female orgasm before the man ejaculates makes conception more, rather than less, difficult, but again little evidence is available on the topic.

Randy Thornhill and Steven Gangestad had a related idea, suggesting that copulatory orgasm is more common when a woman's partner is physically more symmetrical, or similar on each side of an imaginary line drawn down the central axis of an organism. Symmetry is a trait that some argue is linked to fitness in an evolutionary sense, so symmetrical organisms might be expected to show greater survival tendencies. They state that "copulatory orgasm may be designed by selection to promote intimacy (mate selection) with a male of high phenotypic quality" and conclude that female orgasm is a true adaptation, and not a by-product of selection on males.

Whether or not one agrees with the research methods and data analysis of these scientists (and many do not), the part I find troubling is the focus on Where's the Penis. What females experience is still seen as secondary to what happens during intercourse—not before, not after, not as part of a sexual and social relationship. Yes, fertilization and reproduction are the ultimate goals of any behavior from an evolutionary standpoint. But as I discuss in the chapter on homosexuality, even sex is not always about sex. And thus it is unrealistic to expect to define each and every component of sexuality as increasing or decreasing the likelihood of a particular fertilization. I am not saying that sexual behavior lacks adaptations; on the contrary, it is the place where those adaptations may be most crucial. But we must be careful not to make our interpretations too narrow, or to confine them to a male perspective. Hrdy writes, "Yet only a failure to think seriously about females and to consider the evidence would allow someone to conclude that natural selection operates more powerfully on male sexuality than on female sexuality, or to believe that the female's reproductive character could be 'invisible' to natural selection" (1981, p. 173).

IS IT GOOD FOR OTHER SPECIES, TOO?

One of the ways in which many scientists have attempted to make the study of sexuality and of female orgasm less narrow, and to get around the problem of cultural and social context, is by examining the behavior of nonhumans. Do other species show the same pattern as humans?

Answering this question has been difficult because determining whether a female, human or not, has unequivocally experienced orgasm is difficult at best, simply because of the lack of the landmark of ejaculation. In a book chapter called "The Evolution of Female Sexual Desire," Kim Wallen wistfully remarks that "the notion of the sexually passive female is, one hopes, dead," but acknowledges that female sexual desire is a hard thing to evaluate. Its two identifying characteristics in animals do not exactly sound erotic; they are "cooperation" (which to me brings back memories of being in second grade when the teacher asked you to erase the board) and "immobilization," or staying still so the male can get everything aligned properly. Wallen then goes on, "The most striking case is in pigs, where placing a boar's saliva on the snout of an estrous sow produces an immobilization reaction so strong that a human can sit astride her without producing an escape response." While this certainly seems as if it would be an endless source of amusement in research facilities, we are left not really knowing what females feel. According to Wallen, "It is unclear whether female rodents find intromission and ejaculation rewarding or aversive or some combination of both" (p. 61).

Most research on nonhuman female orgasm has focused on primates, for which numerous scientists have claimed to detect movements, postures, or other indications of climax during sex. Macaques, particularly stumptail macaques *(Macaca arctoides),* have received perhaps the most attention in this regard, partly because they are relatively easy to observe under semi-natural conditions of captivity in large enclosures. A female may show a characteristic "clutching" response during copulation, in which she looks back at her partner (these monkeys do not mate face to face, but with the male mounting the female from behind) and reaches to touch him. Taken together with laboratory data obtained from restrained females monitored during clitoral and vaginal stimulation, the behavioral observations suggest that females are experiencing something at least similar to orgasm in human females. In a recent example, Italian researchers Alfonso Troisi and Monica Carosi observed the orgasmlike responses in 80 out of 240 copulations, and found that they were most likely to occur when a female of low social rank mated with a male who was of high rank. In addition, copulations that lasted longer were more likely to produce orgasm. The scientists conclude that social factors, as well as physiological ones, play a role in female macaque sexual response. The Dutch sexologist Koos Slob and colleagues have used the Masters and Johnson criteria, rather than behavioral data, and pointed out that the clutching response and "climax face" do not always accompany physiological responses such as uterine contractions, but that these responses do indeed occur during sexual intercourse.

Similar and equally well-documented behaviors have been noted in other primate species, although scientists have always been hampered by not knowing what to look for. Biologists Malcolm Potts and Roger Short suggested that some nonhuman species may be more inclined toward orgasm because the clitoris is located closer to the vagina in them. Humans, according to this hypothesis, have the difficulties we do because human babies are large and the pelvis had to accommodate them; both the urethra and the clitoris became further separated from the vagina as a result. This idea, while intriguing, suggests that orgasm should be even more common in nonhuman primates than in humans, a conclusion that at least at present is not borne out by the data.

Baker and Bellis are bolder in their claims, and they are the only ones, so far as I can determine, who have considered in any detail whether orgasm extends to animals other than mammals, particularly primates. They suggest, "The possibility that female birds and reptiles might also experience orgasm during copulation has not so far been investigated. . . . It seems likely that a comparable phenomenon will be found to exist in all animals with internal fertilization via copulation and in which the female has sperm storage organs and exhibits flowback" (p. 49) Flowback is exactly what it sounds like: the emission of seminal fluid after copulation. It is relevant here because it fits in with Baker and Bellis's sperm management ideas; if females control sperm by orgasmic movements of their reproductive tracts, they need somewhere to sequester the sperm as well as some way to remove them.

WHAT ABOUT MEN?

In all of these discussions about adaptive value and origin vs. maintenance of orgasm as an evolutionary holdover, one question never gets asked. That question is: why do men have orgasms? Most people would initially say that the answer is so obvious as not to be worth discussion; men have orgasms to ensure that fertilization will take place and reproduction will occur. It makes a process with some long-term costs—care of offspring in some species, risk of sexually transmitted diseases, attraction of predators— have a big short-term benefit.

But think about it. There are at least two problems with this facile conclusion. First, why should males require this kind of evolutionary prodding, when women are going to have to bear at least an equal, and often a greater, share of the results of mating? Given the high investment made by many females in pregnancy, the discomforts of childbirth, not to mention feeding and protection of young, one would expect women, not men, to have such a strong sex drive that they would start yelling in ecstasy when a man pecks them on the cheek. Why do men need the reinforcement of orgasm while women can reproduce perfectly well lying back and thinking of England? The idea that females, whether in other species or our own, lack a strong sexual drive is falling out of favor. Yes, females are selected to be choosy about mates, but they often mate with many males regardless and let the choice occur after copulation, as I discussed in the chapter on sperm competition. It would be a mistake to assume that frequent copulation is always driven by the male, or that selection has left females unenthusiastic simply because they do not have limited access to mates. Potts and Short, in *Ever Since Adam and Eve,* a book about the evolution of human sexuality, declare that "On average, women have a less intense physical sex drive than men, and are usually aroused more slowly. From an evolutionary perspective, they need to be more cautious, and avoiding the 'quick-fix' male orgasm may be to their advantage" (p. 106). This seems puzzling, even if it could be substantiated that women desire sex less than men; why would taking longer to climax once intercourse has begun help a woman be more cautious in mate choice? Presumably the decision has already been made.

The social context can be tricky here, too. To make an analogy: from an evolutionary standpoint, why do we find eating pleasurable? It seems reasonable to assume that the proximate reward of food shapes the ultimate goal of survival. The fact that many Western women are dieting or have psychological problems about eating does not in the least suggest that males and females are selected to respond to hunger differently. Should we conclude that women have a less intense physical hunger drive than men?

The second reason that we should not be so automatic in our assumption about the function of male orgasm is that many, perhaps most, animals in the world reproduce perfectly effectively without any signs of the same type of sexual climax experienced by humans. For reasons that are never clear, scientists who contemplate the existence of orgasm in nonhuman females generally stop after considering primates and a few domesticated or laboratory mammals. But all sexually reproducing animals have sex, by definition, so why should it only be mammals, or even primates, who need the reward of orgasm to help natural selection along? Why shouldn't snakes, and sparrows, and sea anemones experience orgasm?

Perhaps they do. Part of the problem here is defining animal responses in human terms, and more than that, in human male terms. Masters and Johnson did a great service to the study of sex by providing measurable physiological criteria for orgasm, but those criteria, consisting of things like increased heart rate and uterine contractions, are obviously inappropriate for many animals. It seems meaningless to try and monitor sexual arousal in a spider, much less to try and make all other animals fit into the same mold. Therefore we cannot simply conclude that if orgasm makes sex rewarding for humans, it evolved to ensure that we reproduced. What did our ancestors do, before orgasms existed?

I do not have an answer for why males have orgasms. Perhaps it is an artifact of the vasoconstriction and dilation that accompanies ejaculation, and the fact that it feels good is a bonus, a joke, a cosmic perk to make up for prostate cancer, homicide, or the decline of old age. Perhaps it occurs in some form or another in all animals, maybe both sexes, maybe not. The point is that it is foolish to puzzle over why women can conceive without orgasm, over what possible function this trait should serve in females, when we do not wonder why males evolved the same trait. In many if not most species, males at least appear to sire offspring without it either; Baker and Bellis's speculations to the contrary, I remain dubious about the ecstasy that accompanies the cloacal kiss in birds. Nevertheless, the assumption that the male way is the only way has kept us from asking some obvious questions.

Similarly, I do not know whether human female orgasm is an adaptation or an artifact. I suspect it is at least as much of an adaptation as male orgasm, which may mean that both experienced selection as human sexuality itself was evolving, but that both originated as developmental by-products. In her marvelous ode to the clitoris (and to female orgasms), the science writer Natalie Angier muses that, like the music of Bach, such a perfect organ simply had to be. She says, "The clitoris is an adaptation. It is essential, or at least strongly recommended. It is also versatile, generous, demanding, profound, easy, and enduring. It is a chameleon, capable of changing its meaning to suit prevailing circumstances. Like Bach's music, it can always be reinterpreted and updated."

One hallmark of female orgasm is its relative indistinctness, as I mentioned above, which makes it hard to detect in nonhumans. It can also be hard to detect in humans, as many writers and filmmakers have shown; one of the reasons Meg Ryan's famous orgasm-in-the-restaurant scene in the movie *When Harry Met Sally* worked so well was its believability, and not just because Meg Ryan is a professional actor. Maybe female orgasm is so variable and cryptic because it allows females to deceive males about men's sexual prowess. Only the woman can tell a man if he satisfied her in bed. I am sure others can provide scenarios in which such deception, as much a sign of humanity as language and technology, could be adaptive. All of us, however, would benefit by abandoning the male model; female orgasms can be adaptive without being exactly like male orgasms. In other words, we should stop trying to play soccer in bed.

Ten

SACRED OR CELLULAR

The Meaning of Menstruation

IN WHAT HAS TO BE ONE of the most amusing uses of "man" as a false generic, Colin Finn begins a 1987 article on the function of menstruation with, "The phenomenon of menstruation must have puzzled man since time immemorial." Just nine years later, mere puzzlement seems to have escalated; the first sentence of a paper Finn published in 1996 declares, "The significance of menstruation cannot be overstated and questions about its function have worried man since early times." Of course man— or men—may be concerned about menstruation, but it is women who actually do it, as Finn makes clear later in the third paragraph: "The uniqueness of menstruation to woman (and a few other primates) may not have been very apparent to early man." So there we have it: women have been getting their periods forever, but it takes a man to show the curiosity that will tell us what it means.

This chapter is about the puzzle, to all of us, male and female, of menstruation, its evolutionary significance and its meaning for our stereotypes about gender. No one can deny that menstruation is natural, in that it occurs in virtually all women between menarche and menopause when they are not pregnant or lactating. You do not have to go to school to learn how to menstruate, its basic pattern (though not its frequency of occurrence) is seen in women of all cultures, and a woman would menstruate even if she had never seen another woman doing so or read a single tampon ad. Nevertheless, the social associations of menstruation are astonishingly prolific, and it is another subject for which the scientific interpretation becomes bound up with a political one.

Many societies, ancient and modern, say that menstruating women are "unclean," and that they should be isolated to greater or lesser degrees from their usual activities. In some cultures, women go to menstrual huts where they sleep and eat apart from their families. In others, menstruating women are not allowed to prepare food, at least for men, and their presence is thought to interfere with hunting. This last taboo may be related to a still-persistent myth I have heard from hikers, that menstruating women attract the attention of bears, though why bears, which are omnivores and do not usually prey on humans in the first place, should be at all interested by a few milliliters of blood and cellular debris, even if they are able to detect it, has never been clear to me. The U.S. Forest Service went so far as to investigate this claim (behind which there is an interesting tale in itself, I'll wager) and found no evidence for it. Other taboos include menstruating women being prevented from touching the logs or coals from the fires of women who are not menstruating, and being barred from participating in funeral rites or touching of the dead. Apparently this latter tradition, found among the Beng people of the Ivory Coast of Africa, is intended to protect not the spirit of the departed but the woman herself, who runs the risk of perpetual menstruation by contact with the corpse. According to the Gimi of the Eastern Highlands of Papua New Guinea, menstruating women wreak havoc by coming in contact with ordinary objects, and any item thus tainted will break, malfunction, or, in the case of soil where the woman has walked, become infertile and unable to bear crops. The Roman historian Pliny wrote rather forcefully about the destructive effects of menstruation: "But nothing could easily be found that is more remarkable than the monthly flux of women. Contact with it turns new wine sour, crops touched by it become barren, grafts die, seeds in gardens are dried up, the fruits of trees fall off, the bright surface of mirrors in which it is merely reflected is dimmed, the edge of steel and the gleam of ivory are dulled, hives of bees die, even bronze and iron are at once seized by rust, and a horrible smell fills the air."

Sexual restrictions during menstruation are also rampant, mainly those prohibiting intercourse, although some societies have believed that this was the time of the month when a woman could become pregnant. Many anthropologists have weighed in with their theories about the significance of these rituals and proscriptions, with complicated models about moon symbolism, the relationship between blood and fire, and shamans. My point here is that menstruation is viewed as a Big Deal, and not a particularly fun big deal, by many human cultures.

The medical establishment has not been a lot more upbeat about the process. Writing in the early seventh century, Isidore of Seville proposed that: "The menstrual flow is a woman's superfluous blood. . . . On contact with this gore, crops do not germinate, wine goes sour, grasses die, trees lose their fruit, iron is corrupted by rust, copper is blackened. Should dogs eat any of it, they go mad. Even bituminous glue, which is dissolved neither by iron nor by waters, polluted by this gore, falls apart by itself."

I keep wondering where a dog would get enough menstrual fluid to eat, but that aside, the resemblance to Pliny is clear, although Pliny seems to have ignored the problems with glue. Later physicians were no more positive; as the president of the American Gynecology Society gloomily pointed out in 1900: "Many a young life is battered and forever crippled on the breakers of puberty; if it crosses these unharmed and is not dashed to pieces on the rock of childbirth, it may still ground on the ever recurring shallows of menstruation" (quoted in Ehrenreich and English, *For Her Own Good,* p. 110).

Basically, physicians, like the rest of society, saw menstruation as a disease, a defect, something connected with spoilage and fraught with pain and danger for the woman experiencing it. Not all early works contained this portrayal, and a few suggested some positive value to menstruation as a way for women to rid their bodies of impurities, but these were swamped in a flood of medical pejorative. Even texts that were simply describing the process used astonishingly negative terms, including "failed conception," a "weeping womb," and, in a 1913 medical textbook, a "severe, devastating, periodic action" that leaves behind "a ragged wreck of tissue, torn glands, ruptured vessels, jagged edges of stroma, and masses of blood corpuscles, which it would seem hardly possible to heal satisfactorily without the aid of surgical treatment." Although this view has been toned down somewhat over the years, the imagery of menstruation continues to contain words such as "ceasing," "dying," "loss," and "denuding." Carol Tavris, in her book *The Mismeasure of Woman,* and Emily Martin, in an earlier and equally deft analysis titled *The Woman in the Body,* point out that although many of us might think that there is no alternative way to describe the process, in fact the language used to describe similar processes such as the repeated sloughing of the stomach and intestinal lining is very different. Texts use words like "growth," "regeneration," and "renewal," language that puts the event in a much more positive light. Similarly, I have yet to see anyone refer to sweating as the penalty we pay for regulating our internal body temperature; instead, sweat glands, however inconvenient their product, are at least scientifically recognized as part of a fine-tuned adaptive system that allows us to live in a wide variety of climates. Physiology texts do not exhibit distaste for perspiration.

In a 1999 book called *The Curse,* journalist Karen Houppert wonders why modern American society is so obsessed with hiding menstruation from everyone, including the woman who is menstruating. She questions the need we all seem to feel for extreme secrecy, as if being discovered to menstruate were the most embarrassing revelation that could ever occur. Advertising for tampons and pads stresses the confidentiality of the product, featuring models clad in white, and most young men would probably be at least as embarrassed to be asked to buy tampons as hard-core pornography. The word "menstruation" is virtually never mentioned in the commercials or on the packages, and we operate in a world of euphemisms the Victorians would be proud of; women do not bleed, or even menstruate, they have "flow," for which they seek "protection" from the "feminine hygiene" aisles at the supermarket. Houppert interviewed pre-adolescent girls at a summer camp about their attitudes toward menstruation, all of whom were filled with guilty pleasure at the opportunity to utter the words "period" and "blood" aloud. They had a remarkable number of misconceptions about the process, and were desperate for information about something they had begun to realize was to be kept secret at all costs. The book also describes with rather edgy fascination a unique place: the Museum of Menstruation, both a web site and real location, the brain child (or something) of Harry Finley of New Carrollton, Maryland. The museum documents such menstrual milestones as 1955 "sanitary panties" and articles about menstrual fact and fiction.

PERIOD PROBLEM

Maybe this is just one of those strange things about the American historical past and current culture, along with a love of fried food, oddly named teen bands, and stupid jokes. I am not saying that it would be better if we kept nothing private; the magazine ads for foot fungus treatment that show the afflicted digits make me a little queasy, and I am not a big fan of toilet humor on television or in movies. But does it really matter if we discuss menstruation openly, or if medical texts call it degeneration rather than renewal? I think it does, for two reasons.

The first is that creating secrecy and constructing pathology also make a problem for women. On a practical level, Houppert believes that not wanting to talk about menstruation or menstrual products has made it easy for the companies that make tampons to ignore the controversy and potential dangers associated with the dioxin that most if not all tampons contain. It is, she contends, a scandal that "consumers can read a list of ingredients on a shampoo bottle but not on a package of tampons, which are held for hours in one of the most porous and absorbent parts of a woman's body." Periods, however, are not a popular battle cry for consumer activists.

On a deeper level, the clandestine way in which we treat something all women experience quite often has another effect. If something is viewed as a problem, a degrading event full of trauma that is best hidden away, but if we nonetheless all have it, women are left in the same bind as they were by the Broverman experiment on mental health: you can be normal, or you can be feminine, but you cannot be both. An article describing the history of how menstruation was explained by various experts states that 'to an early natural historian, it "was clear that the menstrual fluid was not normal blood." If it is not normal, does this mean that women produce something abnormal, month in and month out, their entire reproductive lives? Presumably the author meant to distinguish between menstrual fluid and the blood in the circulatory system, but the wording is telling. Houppert asks, "What does it mean for a girl, or woman, to say simply, 'This happens to me' and for society to say 'No, it doesn't.' . . . After a while, it becomes psychologically disorienting to look out at a world where your reality does not exist" (p. 9).

The second problem is that such a biased view prevents us from seeing menstruation as one of the great unsolved mysteries in our primate heritage. If, like other aspects of our physiology, menstruation is the result of evolution, what selective forces were important in shaping it? Is the flow itself an adaptation, or is it a by-product of selection for some other process? Do other animals menstruate? What causes variation in the process, and can we use an evolutionary perspective to help us treat those variants that women find unpleasant?

I find it nothing short of extraordinary that until the last few years, absolutely no one offered an ultimate (meaning evolutionary) explanation for menstruation, even though the literature was filled with speculation on the adaptive significance of penis size, sperm length, breast asymmetry, and (lest you think obsession with sex is the only thing governing our interest) the size of different parts of the digestive tract and variation in blood pressure relative to salt intake. Menstruation was just this unhappy inconvenience that women lived with; Gloria Steinem joked that if men menstruated they would brag about "how long and how much," but by and large there was general agreement that, as one scientific paper states, menstruation is "the *penalty* women pay for a greater state of readiness of embryos in their uteri" (emphasis mine). Why look for the selective value of a punishment?

ENTER PROFET

And then, in 1993, a scientist named Margie Profet published a paper called "Menstruation as a Defense against Pathogens Transported by Sperm." In it, she proposed that humans and a few other primates have copious menstrual flows to combat bacteria and other disease-causing microorganisms that may be transmitted during sex. She too evinced some surprise that no other evolutionary explanations had been offered, and included a detailed analysis of what she termed the adaptive design of the endometrium, or lining of the uterus, as well as the constituents of the menstrual fluid, which include some immunologically active cells. She reasoned that anything as costly as the monthly loss of iron-rich blood and other materials that the body has to produce anew with each cycle must have some value to the individual, or it would have been eliminated by natural selection. She then undertook a comparative analysis of where among mammals one might expect to see more abundant vaginal bleeding, suggesting that species in which females have more sexual partners should have more of a need for protection from sperm-borne pathogens. She also predicted that menstrual flow of some kind will be found in more species than are currently known to have it (at the moment the list is limited to many but not all monkey and ape species as well as some shrews). Finally, the paper speculated that if menstruation helps rid the body of infectious agents, giving drugs or other treatments that inhibit bleeding is probably not indicated in many situations.

The work received a great deal of media attention, in itself an encouraging sign in my opinion. But reaction was often polarized; you either loved the idea, because it made a previously taboo and unpleasant process seem valuable, not to mention downright handy, or you hated it, because it was a wild notion postulating a benefit to something that at best people were willing to admit was a by-product. One critic said that "it is very doubtful whether menstruation should be viewed as a separate process. Rather it should be considered as part of the implantation process. . . ." This is an odd interpretation, given that fertilization and implantation themselves could simply be considered part of the process of pregnancy. It also suggests that one just cannot study any physiological event in isolation, a defeatist attitude at best and one that is contradicted by a great deal of medical research. Furthermore, lumping implantation of the embryo with menstruation prevents us from asking questions about why humans seem to have such a generous menstrual flow compared with other mammals, and why overt menstrual bleeding seems to be confined to the primates and shrews. Another scientist flat out accused Profet of putting forth her theory solely for the purpose of defying those negative taboos, rather than because it seemed like a good idea with a basis in scientific fact. Some of the criticisms from the scientific and medical establishment were more thoughtful, and I discuss those below.

IF IT'S NOT A CURSE, IT MUST BE A BLESSING

Many of those on the positive side were part of what Houppert calls the Celebrate-Your-Cycles movement. In books with titles like *Songs of Bleeding, Red Flower, Blessings of the Blood,* and *Dragontime Magic and Mystery of Menstruation,* menstruation is embraced as signaling a connection to the spiritual side of womanhood. Like some of the ecofeminists I discussed in an earlier chapter, the authors of these books responded to the negative biases of society by defiantly celebrating the process of menstruation. Lara Owen, in *Her Blood Is Gold,* suggests, "It is likely that male envy of the blood released by women has fueled many of the negative taboos around menstruation" (p. 38). She wants to hearken back to a time when societies were matriarchal and menstruating women were powerful goddess-types. Other writers advocate rituals for girls at menarche, retreats for menstruating women at which they discuss their emotions and construct altars and decorative "menstrual crowns," and bleed directly onto the ground (preferably covered with moss, understandably enough). Some of the works also romanticize the menstrual huts found in several cultures as places where women, far from being ostracized, are allowed to rest, rejuvenate, and perhaps come up with new weaving patterns, sort of pre-industrial health spas. Profet's theory about the function of menstruation was, as you might imagine, embraced enthusiastically by the cycle celebrants because it supported the idea that menstruation was nature's way of helping us stay healthy.

Arguing against a "technology of suppression," Owen urges women to use pads instead of tampons so they can experience their blood (preferably cloth pads afterward soaked in water, which is then poured onto plants in the garden), avoid painkillers and vaginal deodorants, and increase their awareness of the physical and psychological manifestations of their periods. In Owen's own life, this took the form of apparent negotiations between the author and her cycles: "Maybe my periods wanted to have a more central place in my life and in my awareness. . . . If my bleeding started at the weekend, I stopped driving and stayed home, relaxing in my garden. I remember that it was summer and I lay in the sunshine, just experiencing my bleeding. It was interesting to me—I felt so much better if I just lay about and did nothing" (p. 62).

My reaction to this is, honey, don't we all? But I have no problem with people wanting to water their yards with menstrual fluid (apparently this resulted in a great yield of tomatoes for one intrepid practitioner). And I applaud the idea that girls would find reaching menarche attractive, or at least not horribly embarrassing.

My objection, nonetheless, is the same one I have been promoting all along: it does not help to view science as a tool for ideology, as a weapon in the gender wars no matter which side you are on. Arguing that Profet must be right because a misogynistic society has kept women from appreciating themselves and their bodies is just as foolish as arguing that she must be wrong because she dislikes the fact that menstruation has been such a neglected topic. Both sides ignore what is actually happening, whether support exists for Profet's idea or whether other, perhaps more accurate, theories might be proposed. Taking an ideological stance prevents you from asking what may even be more interesting questions than the one originally posed, and may even lead to some erroneous practical implications.

AN ALTERNATIVE THEORY, AND SOME MYTHS DEBUNKED

In addition to media hype and criticism, Profet's theory also inspired another scientist, Beverly Strassmann from the University of Michigan, to take another look at the topic. Three years after Profet's paper appeared,

Strassmann published one in the same journal, called "The Evolution of Endometrial Cycles and Menstruation." She examined the predictions and assumptions Profet made, and while acknowledging the insight Profet displayed in analyzing the subject to begin with, came up with several counterarguments to those Profet had suggested.

Strassmann said that Profet may have been focusing on the wrong puzzle. The question is not why, if menstruation is costly, it occurs at all, but what the cost would be of not doing it, of keeping the endometrium in a perpetual state of readiness for implantation of the embryo. The lining of the uterus doesn't just sit there like a pillow waiting for the embryo to settle down like a napping cat; it is richly vascularized and contains energy-rich tissue that requires a complex variety of biochemical processes to keep it in shape. These processes cause variation in the woman's metabolic rate at different times of the menstrual cycle. Strassmann calculated that maintaining this physiologically active tissue was quite costly; during the follicular phase, while the egg is being released, metabolic rate averaged 7 percent lower than during the luteal phase, when the endometrium is at its peak. Translated into the amount of food a woman requires daily, this means that losing that demanding uterine lining saves nearly six days' worth of nourishment over the course of four menstrual cycles. "Thus," she states, "the menstrual cycle revs up and revs down, economizing on the energy costs of reproduction." It's not that bleeding is good for you, it's that not bleeding, and keeping the uterine lining intact and perpetually ready for implantation, isn't, at least in terms of energy expenditure.

Strassmann and others also examined the evidence for the pathogen-fighting effects of menstrual fluid, and found rather little. Menstrual blood, however praised or vilified for its other qualities, has bacteria-promoting, rather than eliminating, properties, and there does not appear to be a good relationship between the number of mates or sexual partners common in a given species and the likelihood of its exhibiting menstruation. Endometrium that does not receive a fertilized egg is reabsorbed by the body in many mammal species, but in humans and some others, the endometrial tissue appears to be too abundant for this to occur, although approximately two-thirds of it is taken back up into the body rather than shed. Finally, the iron lost during menstruation is likely to be minimal and relatively easily replaced if a woman's diet is adequate.

Along the way, Strassmann took a swipe at another bit of menstrual lore that may be turning out to be false: menstrual synchrony, the supposed convergence of women's menstrual cycles onto the same pattern when they are in close quarters. The original work documenting this phenomenon was done by the eminent psychologist Martha McClintock when she was a senior in college and studied her dorm-mates; she published this honors thesis in 1971 in the prestigious scientific journal *Nature*. The notion proved remarkably appealing, and the idea that sisters and other groups of women develop bonds, not just emotionally, but in their physiology, became conventional wisdom among scientists and nonscientists alike. Several evolutionary biologists came up with theories to explain why menstrual synchrony would be adaptive in pre-industrial societies where most of the women in a tribe or band would be expected to be fertile at the same time.

The problem is that such synchrony may not exist. The original study and a few later ones claiming to substantiate McClintock's findings suffered from some statistical problems, including a failure to correct for the likelihood of women's cycles coinciding through chance alone (a larger probability than you might think—remember that the chance of two people having the same birthday in a group of only fifty is a whopping 97 percent). In addition, the studies did not use subjects from populations with natural patterns of fertility, minus the cycle-altering effects of the birth control pill.

Strassmann studied the Dogon, a farming people in Mali, West Africa. She monitored their reproductive cycles through native informants and data collected from urine samples of over one hundred women for more than two years. She found no evidence for synchrony, and furthermore, no support for the idea, long touted by cycle celebrants and others, that menstrual cycles are governed by lunar ones. The Dogon lack electric lights, so if anyone is going to show an influence of the moon on biological rhythms, they should, but Strassmann found no such thing. This is not to say that biological clocks and circadian rhythms do not occur, just that the popularized effects of the moon on women appear to be dubious at best. The jury is still out on whether menstrual synchrony exists in some circumstances but not others; again, the point is that accepting an idea simply because it sits well with an ideology is risky.

MENSTRUATION AND CONFLICT

The other cultural icon of menstruation that Strassmann examined was the menstrual hut. In 12 percent of pre-industrial societies, women are segregated in separate areas or structures during menstruation. Among the

Dogon, women sleep in the huts for five nights during their menses and cannot go into the streets of the village or cook for their husbands; sexual intercourse is forbidden. The belief that women are most fertile immediately after menstruation is widespread in the society, and Strassmann suggested that instead of being a haven from daily activity and a place to renew connection with the earth, menstrual huts allow Dogon men to monitor the reproductive status of women, and thereby prevent women from deceiving men about the time when they conceived. This provides a check on adultery because a woman cannot claim that she became pregnant months earlier or later than she did by asserting that she had menstruated since a suspected sexual encounter.

Using hormone levels in urine samples collected from the women throughout their cycles, Strassmann determined that few if any of them lied about their menstrual status. Furthermore, the menstrual-hut-as-resort hypothesis was belied both by her informants, who told her that women did not want to go to the huts, and by the discovery that whether women used the huts depended on their husbands' religious beliefs, but not their own. The wives of husbands who believed in animism used the huts; those of nonanimist husbands did not. This suggests that male interests are served by the advertisement of reproductive state, while female interests benefit from masking signs of the fertile period. The threat of both social and supernatural sanctions apparently keep the Dogon women from breaking the taboos. A woman married to an animist who does not go to the hut when she menstruates suffers a fine of one sheep, which the male elders sacrifice and eat. Life in the huts is reputedly confining and dull, and women are still expected to work in the fields when they are menstruating.

Conflicts of interest between males and females over reproduction are common, as I have mentioned repeatedly. Females are better off when males contribute to the care of their offspring, but males benefit only when they care for offspring that are genetically related to them. Ovulation, of course, does not occur just after menstruation but midway through the cycle, and therefore the Dogon are wrong about when women are most likely to become pregnant. Knowing when a woman menstruates helps, but it is not the entire story. Just as with menstrual taboos, theories about the significance of this concealed ovulation abound among anthropologists and evolutionary biologists, but it is possible that, evolutionarily speaking, the hidden nature of this most fertile time may be the Dogon's—and perhaps women's in general—way of having the last laugh.

The Dogon are not the only people with menstrual huts, and they are certainly not the only people with menstrual taboos. But Strassmann's analysis of their culture points to what I see as a more productive line of inquiry than either ignoring the social context of menstruation and focusing solely on the physiology, or else divorcing the process from its biological significance and suggesting that social taboos associated with menstruation function to make women feel ashamed about their bodies and bodily functions (this may indeed be an effect, but it is likely not the context in which the behaviors evolved).

Another conflict has been suggested to figure in menstruation, but here the struggle lies not between men and women, but between the woman and a newly fertilized egg. As I discussed in the chapter on motherhood, the idea that women are always going to put the interests of their offspring before their own is filled with caveats, with "yes, but . . ." and "assuming that . . ." qualifiers. Put the children first, unless your reproductive interests in the long term are better served by killing, starving, or abandoning them; in most situations, of course, that is unlikely to be the case, but the conflict is there nonetheless. David Haig, a biologist at Harvard University, has gone so far as to suggest that the battle begins even before birth, as the developing fetus seeks the maximum amount of nutrients from its mother while the mother benefits by balancing the needs of the fetus against the preservation of her own ability to reproduce in the future. Carried to its extreme, this idea predicts that on occasion, a very young embryo will survive better if it "hides out" in the uterus and implants even though doing so is against the reproductive interests of the woman carrying it. Or it may simply be defective and unlikely to warrant the investment that a full-term pregnancy requires. Presumably once the embryo gets to a certain stage of development it becomes more likely to be retained. Alternatively, if the woman's body detects it in time, she can remove the fertilized egg by shedding part of the endometrium. Haig reasons that menstruation could serve to flush out the offenders more effectively than partial endometrial shedding because it eliminates the entire lining of the uterus and affords no places for the embryo to sequester itself. He points out, "Sloughing the endometrium is an effective means of eliminating a single embryo but would be an indiscriminate form of embryo selection if it resulted in the loss of an entire litter. Menstruating species usually produce singletons" (p. 81).

I am not altogether won over by this idea that menstruation is a tool in parent-offspring conflict. The problem is not the implication that a few-celled zygote can have a behavioral strategy; as I have mentioned before, there is no need to invoke conscious decision-making in animals when they fight, mate, or play, so long as behaving in a certain way enables an individual to gain a reproductive advantage over individuals who behave otherwise and to have offspring that are more likely to behave as did their parents. Analogies about warfare and battles are rife in the scientific literature about the immune system ("killer T cells" is but one example), but no one really expects that there are miniature General Pattons amidst the white blood cells. My reservation comes more from the extremely limited nature of menstruation among animal species; menstruating species may produce one young at a time, but so do many nonmenstruating species, and it seems to me that other possibilities exist for the elimination of disadvantageous embryos. But the facts are by no means all in, and ideas like Haig's are at least potentially testable now that the notion of menstruation as a phenomenon requiring explanation has been put forward.

TOO MUCH OF A GOOD THING?

If menstruation is natural, how often should we be doing it? This question is less facile than it might appear. In a modern Western society, where women do not usually marry near menarche and generally limit the size of their families far below what would occur without birth control, women can have a great many menstrual periods in their lifetimes. Pregnancy and lactational amenorrhea (the ceasing of ovulation and menstruation while a woman is nursing) reduce the number by a great deal in some other societies, but Western women are infrequently pregnant and often curtail nursing a few months after giving birth. Although one can use a variety of methods to calculate the numbers of cycles in theory, surprisingly few studies have done so empirically, by simply keeping track of the number of menstrual cycles real women experience. An American physician did so with her own cycles; with time out for giving birth to three children, she had a total of 355 menstrual cycles in her lifetime. She was reproductive for thirty-two years, while thirty-eight years appears to be more typical for American women, so this number may be an underestimate. Another study came up with 450 cycles, which may be on the high side. It is undeniable, however, that either number is a great deal higher than that of women in cultures without birth control. The ever-patient Dogon were also tallied in an effort to address this question, and Strassmann calculated that the average number of lifetime menses was 128, or about a quarter to a third of the number experienced by women in many industrialized societies.

Other factors, such as nutrition during childhood and the incidence of diseases that affect the time of menopause, will obviously influence these numbers. The point is that although it is undeniable that menstruation is "natural," that does not mean it is desirable to do it as often as many Western women seem to. Evidence is accumulating that some reproductive cancers and other diseases may be found at higher levels in women from industrial societies than in those from pre-industrial societies because their bodies are exposed to higher levels of certain hormones than would be expected to occur were they pregnant and nursing more frequently. This does not suggest that pregnancy is advocated as a cancer preventative, but it may mean that birth control pills or other mechanisms for reducing the number of periods women have in their lifetimes could have an added benefit of reducing some gynecological diseases. Rather than being negative about suppressing something women are supposed to do, it behooves us to think about whether we really are supposed to do it. It is certainly too early to draw conclusions that are applicable to physicians making clinical recommendations. But consideration of what evolution has suited women's bodies to do has a place in medical science. (Those interested in the broader implications of this viewpoint should consult the growing literature on Darwinian medicine.) This to me is a more realistic way of accepting menstruation as a part of our lives than either assuming it is at best a nuisance and at worst a pathology, or else championing monthly bleeding as a way to feel spiritual sisterhood.

Don't get me wrong; I like the in-your-face defiance of the women who point out that it is ridiculous to act as if menstruation were something to be embarrassed about, hidden at all costs, especially from men. The article that eventually led to Houppert's book was published in the *Village Voice,* where it had a cover photo like one on the book itself, a photo that (as Houppert herself puts it), "looked like any of a dozen provocative ads for skin creams, perfumes, or health clubs: a woman's sexy lower torso in profile, smooth thighs and pert butt alluringly displayed. But here, peeking out from between the woman's thighs, was a tampon string" (p. 8).

The reaction was vitriolic, worthy of a display of hard-core porn, not a simple piece of cotton. Letters to the editor were written, controversy raged, disgust and outrage were expressed. It seems to me that our society exposes itself here as a wee bit on the irrational side about a biological process. I hope that in a decade or two our horror will look as silly as the Victorian reluctance to say the word "leg" about the supports of a piano. But the solution is not to decide that we will promote one interpretation of the significance of menstruation over another simply because of politics. Rather, it is to try as best we can to understand the process, in all its physiological and evolutionary significances.

Eleven

THAT'S NOT SEX, THEY'RE JUST GLAD TO SEE EACH OTHER

THERE IS A "GAY GENE." There is not a "gay gene." Homosexuals are born that way, homosexuals choose to be that way, we are all basically bisexual but societal pressure forces most of us to choose a single polar sexual orientation. It is natural, so we cannot blame those who are homosexual because they simply are born to be sexually attracted to members of the same sex. It is natural, but it is sick, and so we should love the sinner but hate the sin. It is not natural, and those who are homosexual are exhibiting an aberration, a pathology, and need help in choosing a better lifestyle.

The causes and origins of homosexuality are much in the news these days. Biologists are called upon to testify in court cases, geneticists appear on the television show *Nightline,* the reprints of a neurobiologist are available to anyone who calls the toll-free number of a conservative antigay group. Books are titled *Queer Science, Straight Science,* and *The Science of Desire.* And when it comes to animals, homosexuality seems to defy the basic precept of biology: sex is for reproduction. How could behavior evolve if it does not contribute to the reproductive success of the individual?

Homosexuality, then, is the perfect place to see what happens when our ideas about nature, gender, culture, and the differences between humans and other animals collide. In a recent book called *Biological Exuberance,* Bruce Bagemihl claims that sex between males and between females is common among animals, that such activity has a function, and that scientists have spent the last many years ignoring or explaining away these behaviors. In this chapter I examine the basis for thinking that homosexual orientation in humans has a biological basis, meaning that it is an inherent trait like being right- or left-handed, and then see how this information does and does not help us understand homosexuality in animals. This perspective is the reverse of the usual process, that is, seeing how a trait in animals can shed light on the same or similar characteristics of humans, but I think it is a potentially useful one, and may help us understand the role of sexual behavior itself.

GAY BRAINS, GAY BROTHERS, GAY GENES

Several lines of research have suggested that people, particularly men, who are sexually attracted exclusively or nearly so to the same sex share certain anatomical or genetic characteristics. First was the headline-generating study of Simon LeVay, a neurobiologist from the Salk Institute who examined the brains of nineteen gay men who had died of AIDS, sixteen heterosexual men, six of whom had died of AIDS, and six heterosexual women. He was particularly interested in comparing the size of a region called INAH, the interstitial nuclei of the anterior hypothalamus (nucleus in this context refers to a collection of nerve cells). The hypothalamus is a small region at the base of the brain known to be associated with several essential activities, including eating and drinking as well as aspects of sexual behavior, and it is found in some form or another in all vertebrates. This particular part of the hypothalamus was found to be larger in men than women, and LeVay and his colleagues, including Laura Allen, thought it would be worth seeing if INAH also differed depending on sexual orientation. It did; in fact, one of the nuclei, INAH-3, was twice as large in the heterosexual men as it was in the homosexual men. LeVay wasn't necessarily saying the men were "born that way," just that a difference in a part of the brain that should be relevant to sexuality existed between them.

Controversy was not long in coming. Aside from opposition on purely political grounds, some scientists challenged LeVay's findings by pointing to the difficulties of delineating parts of the brain with the precision he claims to have accomplished. "Reading" the brain is like reading any other kind of map; one needs manufactured landmarks and color-coding. Brains do not come marked with the boundaries of the hypothalamus and cerebrum any more than one can drive to black lines of latitude and longitude or see on the ground the border of Tennessee. To demarcate areas on the brain, scientists make extremely thin slices of tissue, called sections, which are mounted on glass slides and—this is the crucial part—stained with chemicals that are differentially absorbed in different places. A few neurobiologists criticized the LeVay lab's use of one of these stains in particular, claiming that it does not yield the desired effect in all cases. Others questioned the initial finding of the extreme sexual dimorphism of the INAH, and yet others were concerned that using AIDS victims compromised the results because changes in the brain could be an incidental side effect of medication or of the disease itself. Furthermore, much of the basic information on how we think the brain works, particularly with respect to sexual behavior, comes from the favorite model species, laboratory rats, and as I noted earlier, generalizations based on one or a few animals evolved in their own environments can be risky. The discussions continue, and at the moment some differences between the brain tissue of homosexual and heterosexual men appear to exist, though the interpretation of those differences remains unsettled.

The second set of studies pointing toward a biological basis for homosexuality were what are called, logically enough, twin studies. As everyone who has sat through an introductory biology class knows, twins come in two types: identical, more properly referred to as monozygotic, and fraternal, or dizygotic. The zygote is the fertilized egg cell that eventually develops into an embryo, and if a single zygote splits, the genetic material in each part is virtually identical and you end up with the kind of twins portrayed in *The Parent Trap,* arising from the same egg. If two zygotes— different eggs, different sperm—both implant in a woman's uterus and develop into viable fetuses, you end up with offspring that may resemble each other as much or as little as any other siblings, like Arnold Schwarzenegger and Danny DeVito in the movie *Twins.* Scientists are fond of studying twins because they allow at least a partial teasing apart of genetic versus environmental factors in some traits, and monozygotic twins share a remarkable number of characteristics, including behavioral ones, even when they have been reared apart. Note that even monozygotic twins are not truly genetically identical; they differ in the DNA they inherit, not from the chromosomes, but from other parts of the cell called mitochondria. Mitochondrial DNA is always maternally inherited, rather than being half from the mother and half from the father, and twins do not get the same mitochondrial DNA. Other not-strictly-genetic differences may arise in the uterus, where subtle variations in fetal position may be influential in later life.

These caveats aside, however, monozygotic and dizygotic twins make good comparisons because you do not expect the latter to differ much more or less than other pairs of siblings, while the monozygotic siblings should be very similar, at least in traits with a large genetic component. So psychologist Michael Bailey and psychiatrist Richard Pillard compared the incidence of homosexuality in fifty-six monozygotic twin brothers and fifty-four dizygotic twin brothers, as well as in fifty-seven adopted (and therefore genetically unrelated) brothers. The subjects were volunteers who answered ads recruiting them. It turned out that if one twin was gay the other one was also gay in 52 percent of the monozygotic pairs but only 22 percent of the dizygotic pairs and 11 percent of the adopted brothers. A similar study with lesbians gave comparable results. Bailey and Pillard concluded that because monozygotic twins are so alike genetically, the higher incidence of shared homosexuality suggested a genetic basis for this trait as well.

Here the criticisms fell into two main camps. First, because the brothers and sisters had all been raised in the same environment, one is left not knowing if the greater level of similarity between the monozygotic twins arose because people treat them differently from their dizygotic counterparts, let alone adopted children; they are more likely to be dressed alike, be mistaken for each other, and get parts in television commercials. How might this influence the likelihood of them both developing the same sexual orientation? The second point is not so much a criticism as an observation, and one with which Bailey and Pillard would probably agree: even if monozygotic twins are more than twice as likely to share the trait of homosexuality as dizygotic twins, there are still plenty of monozygotic pairs out there in which one member is gay but the other is not. Therefore genes cannot possibly be the sole determining factor in sexual preference. How significant this conclusion is depends in large part on one's social and political biases.

The last major brick in the wall is the source of the phrase "gay gene." A geneticist at the National Cancer Institute, associated with the National Institutes of Health, had been researching the regulation of a gene producing a protein that binds to metals such as copper and mercury. But after over twenty years of this work, which had no social implications whatsoever, Dean Hamer decided to make a radical switch, and start studying the possibility of a genetic basis for homosexuality. I was interested to learn from Hamer's book *The Science of Desire*, written with Peter Copeland, that a major impetus for this change in research interest was Charles Darwin's *Descent of Man and Selection in Relation to Sex,* published in 1871 and still a heavily cited classic. In it, as I discussed in the first chapter, Darwin reflects not on the source of diversity among different species and how organisms are related to each other, as he does in the better-known *Origin of Species,* but on the source of differences between males and females of the same species. He coined the term "sexual selection," and also mused on the cause of variation in human behavior in different cultures. I use the book as a text in my graduate behavioral ecology course, and still find its insights fresh after well over a century. What Hamer gleaned from the book was that in the early days of evolutionary theory, the idea that behavior as well as physical characteristics could be inherited was at least as acceptable as the idea that organisms evolved in the first place.

Hamer also read *Not in Our Genes,* a critique of sociobiology studies examining a genetic and evolutionary basis of behavior, written by the Harvard geneticist Richard Lewontin. Hamer disagreed with the political slant Lewontin professes, but the book gave him his first inkling of just how emotionally charged and divisive research into the biological basis of sexual orientation was likely to be. Nevertheless, he and a team of researchers began a study to see, not only if there was a genetic basis for homosexuality, but if they could localize the chromosomal region where the gene or (more likely) genes occurred.

They began by interviewing a large sample of individuals to address two questions. First, what is the distribution of the trait they were trying to study? Depending on the answer, a search for a gene or genes connected with the trait could be either reasonable or completely impossible, even if the trait itself is undeniably genetically controlled. Take for example the traits of handedness, whether a person uses the right or left hand, and height. Both "run in families," and while many left-handed people have been forced to change the hand they use, children start favoring one or the other hand quite early in life, and left-handed children are more likely to have left-handed relatives. Similarly, adult height can be predicted from a child's height at a given age if the parents' heights are known.

In neither case do we know which genes on which chromosomes are responsible for the characteristic, but if a funding agency were to put its money on scientists finding such genes, it would be far more prudent to invest in one than the other, solely on the basis of likelihood of success. The reason is that height is virtually certain to be caused by a great many genes with differing effects in different circumstances and with complex interactions among them. Height has a continuous distribution: adults range in height from, say, 4 feet 6 inches to maybe 7 feet for certain basketball champions, and—this is the crucial point—everything in between. There are human beings who are 4 feet 7, 4 feet 8, 4 feet 9, and so forth, with a bulge in the curve around 5 feet 6 for North Americans, but with representatives from each possible height category. Height is what we call a quantitative trait or a metric trait, with variation that is smeared across a range of values. Such traits may have a genetic basis, and we can determine the proportion of variation in them that likely comes from the parents, but the continuity of that variation arises from many genes with multiple effects. Localizing them with modern molecular techniques would require a Herculean effort, if it could be done at all.

Handedness is different. People, at least in childhood, show extremely strong preferences for the use of one hand over another, and very few are ambidextrous, although one might think that such versatility would be advantageous. Nevertheless, around 90 percent of people are right-handed, and a steady 10 percent are left-handed. Unlike height, handedness is a dichotomous trait, and it is probably controlled by only a handful of genes. (Handedness does show some genetic peculiarities, in that the predilections of parents are not cleanly passed on to children, but that is apart from the distribution of the trait.) No one has undertaken a search for such genes, in large part because it isn't clear why anyone would want to know, but the hypothetical funding agency would have a far better chance of a return on its money if it supported such a study than if it asked scientists to find the genes influencing height.

Hamer therefore reasoned that his work would have a higher probability of success if homosexuality looked like handedness than if it looked like height. On the basis of extensive interviews, his research team classified men's sexual orientation as 1 if they were completely heterosexual and 6 if they were completely homosexual, with the possibility of gradations in between. Somewhat to their surprise, relatively few individuals fell into the middle ground. The interviews were designed to detect, insofar as possible, not variation in behavior so much as variation in orientation— which sex men had sexual fantasies about, to which sex they were attracted even if they did not act on their impulses. Although such techniques can be criticized, they give us the best clues we have about the distribution of the trait, and it looked as though sexual orientation had a discontinuous distribution. So Hamer was prepared to begin his search without fearing that the genetics would be too complex to permit an answer.

He next assembled pedigrees or genealogy charts of all the families with gay members that he had interviewed plus a few from other sources. Here he found another helpful piece of evidence: homosexuality was more likely to be traced through the maternal line than the paternal one, which suggested to him that the X chromosome, which males get from their mothers, was a likely place to start hunting.

Using a computerized database of genetic information, Hamer picked a region of the X to investigate, Xq28. Several months of hard work later, he and his laboratory group were convinced they had something; particular manifestations of markers at the Xq28 region were far more likely to co-occur in homosexual brothers than would be expected by chance. In a 1993 paper, published in the journal *Science,* where LeVay's research had appeared a few years earlier, Hamer and colleagues made the carefully worded claim, "We have now produced evidence that one form of male homosexuality is preferentially transmitted through the maternal side and is genetically linked to chromosomal region Xq28."

Again, the response was thunderous, both pro and con, and again both scientists and the public had their reasons to favor or discount Hamer's findings. Note that Hamer et al.'s statement was not a declaration of a gene that makes people gay, less still of a way to transform people's sexual orientation. Nevertheless, the media seized upon the research with the excitement usually reserved for the peccadilloes of movie stars and presidential candidates. London's *Daily Telegraph* headline read, "Claim that Homosexuality Is Inherited Prompts Fears That Science Could Be Used to Eradicate It," while the *National Enquirer* cut to the chase with, "Simple Injection Will Let Gay Men Turn Straight." Research is continuing, and it is likely that other regions in other chromosomes will be found to influence sexual orientation and that the exact location of the gene in Xq28 will be pinpointed within a few years. Work attempting to replicate that done in Hamer's lab has had mixed results, so the picture is as yet incomplete. Nonetheless, a recent survey of 508 psychiatrists, the same people who not too many years ago classified homosexuality as a distinct mental illness, showed that the explanation most favored for what makes men homosexual was genetic inheritance.

DO BIRDS AND BEES DO IT, TOO?

Where do animals fit into the debate over the origins of homosexuality? One writer states, "Sexuality and culture are the very essence of what makes our species . . . distinctive," and claims that humans are unique because we have "symbolic institutions that create meaning and purpose in life beyond reproduction." Sex without procreation, including sexual behavior between individuals of the same sex, is thus presumably reserved for humans. Conversely, homosexuality has been seen as representing the primitive state, an infantile state of development of the sexual being, which might suggest that it is reasonable to expect it to occur in animals. A religious newsletter points out that homosexuality cannot be viewed as a negative act, "like lying or stealing," because "it occurs in animal species in which these actions—these sins—cannot occur." Presumably the authors would consider kleptoparasitism, the taking of food from one individual by another (a common method of foraging in several species of gulls and their relatives the skuas) to be stealing, but by definition not sinful, if sins only occur in humans, which begs the question of the moral status of any act, including homosexuality, that occurs in people and other animals.

Are nonhumans homosexual? A Florida man was convicted of animal cruelty for killing his wife's neutered male poodle–Yorkshire terrier, which he claimed had sent him into a rage by making sexual overtures to the family male Jack Russell terrier. Popular reactions aside, researchers have known for a long time that experimental manipulation of hormones in rats and other laboratory rodents could induce males to mount other males and females to mount females. And a mutation in the old standby, the fruit fly *Drosophila melanogaster*, is associated with males courting other males instead of females. All lines of flies resulting from mutations are given names to distinguish them from one another, and often the names are whimsical, like the *hni* line (pronounced "honey") of males exhibiting low courtship levels, which stands for He's Not Interested. In a particularly wince-producing example, the male-courting males were first called *fruity*, later changed to *fruitless*. Are the flies gay? Are human gays mutants? At the very least, the mutation suggests that sexual orientation is governed, in fruit flies, by the genes. But we already knew that; no one was suggesting that their sexual orientation developed during larva-hood, perhaps from the actions of a distant father and an overprotective mother.

Simon LeVay points out, "The question of whether animals engage in same-sex sexual behavior has been debated for centuries, most often in the context of efforts to stigmatize homosexuality. Three classes of answers have generally been offered: 'Animals don't do it, therefore it's unnatural'; 'Animals *do* do it, therefore it's bestial'; and 'Some animals do it, and those are the unclean animals'" (p. 195).

Dean Hamer is perplexed by interest in whether nonhumans are homosexual; although a letter he received claimed that "If none of the lower orders engage in sex with the same gender, the motivating factor for homosexuality must not be genetic, rather it must be in the noggin," Hamer is unconvinced that data one way or the other would shed any light on the politics. He states, "Personally, I don't see why people are so interested in what happens in the barnyard; sometimes it seems they are more interested in how animals have sex than how we do." The whole topic is irrelevant to studies of humans, he suggests, since human sexuality is so different from that of animals, and since animals do so many things we do and do not find mirrored in human behavior. Yes for parental care of young, for infanticide, for picking ripe fruit from trees. No for driving, for wearing clothes, for watching movies. Does it really matter which of these lists homosexuality goes onto?

Bruce Bagemihl thinks it matters very much. In *Biological Exuberance,* he claims that animal homosexuality is so ubiquitous that it challenges our ideas about the function of both human and nonhuman sexual behavior. Same-sex behavior, including courtship and parental activities as well as actual copulation, has been noted in many different kinds of animals; Bagemihl documents its occurrence in at least 450 species, but points out that this is probably a vast underestimate both because behavior is generally difficult to observe in many small, nocturnal, or simply uncooperative species and because scientists themselves have been reluctant to acknowledge the appearance of homosexuality in their study system. Even heterosexual mating has not been seen in numerous species known to reproduce sexually. To further complicate matters, in sexually monomorphic species—those in which males and females are virtually identical in appearance, like crows—sex is frequently assigned on the basis of behavior, with the partner doing the mounting in a sexual encounter presumed to be male and the one being mounted presumed to be female. It is therefore difficult to obtain an accurate estimate of homosexual behavior in animals either in the field or in captivity (though the religious newsletter I mentioned above inexplicably declares that it "exists in proven ratios in all mammal species").

Bagemihl includes an exhaustive listing of observations of same-sex behavior in nonhumans, ranging from male butterflies attempting to engage the genitalia of other males in flight to the apparently Dionysian life of bonobos. He cites the apparent long-term homosexual preferences of certain rams both in zoos and on farms as well as in wild bighorn sheep populations, and notes the horror with which the latter was received by the scientist making the observations: "I still cringe at the memory of seeing old D-ram mount S-ram repeatedly . . . to state that the males had evolved a homosexual society was emotionally beyond me. To conceive of these magnificent beasts as 'queers'—Oh God!"

Other authors did not acknowledge their biases so openly, but apparently sexual behavior between males or between females was often dismissed as a case of mistaken identity, a form of aggressive behavior, a way of reducing tension in social groups, or, as reviewer Susan McCarthy phrases her favorite, a Really Big Greeting. Pairings of female gulls, commonly observed nesting together on islands off the California coast, were said to be the last resort for populations in which males were in short supply. If the behavior occurred in captivity, people often used the prison analogy; the animals would not seek pairings with members of the same sex if they were living a more "normal" life. Bagemihl casts doubt on all of these explanations, and makes a convincing case for homosexual behavior being an established part of the lives of many species of animals. McCarthy, writing for the online magazine *Salon.com,* confesses, "There's a certain temptation to leaf through the book shouting 'Caribou? Gay! Red-necked wallaby? Gay! Golden Plover? GAY GAY GAY!' "

Whether one is delighted or dismayed by the evidence, and whether one really wants to accept every instance mentioned by Bagemihl as documenting homosexual behavior to be characteristic of a given species (sometimes same-sex encounters were observed only once during an extensive field study), finding homosexual behavior in animals is significant for several reasons. It is completely meaningless, however, for another reason, and unfortunately that is the reason many people are interested in it in the first place.

NATURE, NURTURE, AND NONSENSE

As I have been arguing throughout this book, using information about animal behavior to justify social or political ideology is wrong. Whether or not animal homosexuality is widespread or occasional, whether it occurs more often in some species than others, should not influence our public policies and decisions about legal protection against discrimination for homosexuals. Hamer's blunt statement which I quoted above is in agreement with this idea, as are virtually all the other commentators on animal homosexuality, Bagemihl included. What many people fail to realize, however, is that the same can be said for the evidence about the biological or genetic basis for homosexuality in humans.

Virtually all the major players in the biology and homosexuality question—Hamer, LeVay, Bailey and Pillard, and even William Byne, one of the most vocal critics of the INAH work—emphatically state that policy should not be based on their findings. Nevertheless, the media, special interest groups (both for and against gay rights), and to a large extent the general public are following the debate to find out whether homosexuals cannot help their sexual orientation. Dean Hamer received a letter from the father of two gay sons who had been told by his church that their homosexuality was caused by poor upbringing, so that the father was at fault. Upon reading of Hamer's work on Xq28, he felt immense relief, because the implication was that his sons' rearing had nothing to do with their sexual orientation. He could accept himself as well as his sons for what they were, naturally. Hamer reflected, "Perhaps I should have been gratified by testimony like this. . . . Instead, I was saddened. This man had changed the course of his life, and the lives of everyone in his family, because of a few paragraphs in a magazine. . . . But what if the experiment had failed . . . or what if . . . the sons were gay for some other reason? Then would this father go back to blaming himself for raising two gay sons, and would they be less worthy of his love?" (p. 19).

Let's face it; Pat Robertson, Jesse Helms, and Jerry Falwell are not going to change their minds about homosexuality no matter how many pairs of lesbian gulls they see, or how many linkage analyses are performed. Similarly, members of gay rights groups are not going to abandon activism against discrimination if it turns out that LeVay misinterpreted his results or Hamer didn't calculate his genetic marker associations correctly. Nor should they. People need to be able to make decisions about their lives without worrying about keeping up with the bonobos. Attempts like the one by the Colorado Supreme Court to call people like Byne and Hamer to testify for or against Amendment 2, the so-called antigay initiative which prevented homosexuals from receiving legal protection against discrimination, are completely misguided. Byne was outraged that his opposition to the interpretation of LeVay's results would be taken to mean that he believed homosexuality was a "lifestyle choice" and hence the choosers were undeserving of civil rights. Furthermore, while some gay activists feel vindicated by a belief that their sexual orientation is "not their fault," others are leery of any use of biology in politics, since biology has been so misused before, particularly in setting women and minorities as biologically, inevitably, inferior.

One reason that the use of the scientific data for such aims is misguided is the point I raised when I discussed motherhood: all behavioral traits, including homosexuality, are both environmentally and genetically determined. Indeed, the same is true for traits in general to a greater or lesser extent. It is the difference between traits which can be said to be learned or innate, which is why the twin studies are particularly relevant. No one would ever suggest that monozygotic twins, whether reared apart or together, would always have the same sexual orientation. The relative difference between both twins being gay versus both liking peanut butter could be determined if the twins had grown up in different (and equally peanut butterless) environments, if anyone would ever be interested in such a thing. But that is a different question.

What, then, are the lessons to learn from the occurrence of homosexuality in animals? I think there are two. First, how did such an apparently disadvantageous behavior come to be? Second, what does homosexuality tell us about the nature of sexual behavior itself?

EVOLUTION AND HOMOSEXUALITY

It is obvious that if homosexuality is even partially inherited, a paradox arises because exclusive homosexuals do not reproduce, and hence their sexual orientation should likewise die out in time. Many solutions have been proposed over the years, including the idea that homosexuals' genes are passed on via their extended families, as by nieces and nephews. If an individual helps a relative sharing a proportion of that individual's genes, and enables that relative to have an even greater reproductive success than the individual would have had if it had reproduced on its own, even seemingly selfless behaviors such as rearing someone else's offspring may make good evolutionary sense. Whether these benefits accrue in humans is unclear, but it is unlikely that they are the governing factor in animal same-sex relations because they occur in species where the opportunities for such help are limited at best. Another suggestion is that truly exclusive homosexuality is confined to humans in recent history. This too seems dubious, and insofar as we can glean, proportions of homosexuals have been remarkably consistent both in history and in different cultures, despite enormous variation in societal acceptance of them.

Yet another possibility is that homosexuals are extraordinarily fecund or have some other survival or reproductive advantage when they do reproduce. A different side of this view is that homosexuality happened along the way while natural selection was happening to something else. People often think that each and every trait must be good for the individual bearing it, because they assume that natural selection has acted on every trait. But a few moments' reflection will generate many characteristics that we would be better off without, including cancer and appendixes that are prone to life-threatening inflammation. Other characteristics seem like neutral variation, but are maintained nevertheless, such as whether or not one can roll one's tongue into a cylinder. Evolutionary biologists have many explanations for such genetic variation in traits, including their persistence in populations despite apparent disadvantages in reproduction. One suggestion is that the genes associated with homosexual behavior are also closely linked to genes enhancing reproductive success on the occasions that homosexuals do engage in heterosexual relations, perhaps by conferring an advantage in sperm competition. A related idea proposes that homosexuality is influenced by a partially recessive allele, so that the same-sex orientation is expressed only when two copies are present in an individual. If only one copy is present, the individual is not homosexual, but experiences some advantage in reproduction that overrides the disadvantage experienced by those not producing offspring when both copies occur. What that advantage might be remains completely obscure, and indeed all of these arguments are for the moment quite speculative. They also rarely address the question of homosexuality in females, whose reproductive success is likely to be less variable than that of men in the first place for reasons discussed in earlier chapters.

It is important to note that whatever the evolutionary advantage of homosexuality might be, it does not occur at the expense of the individual for the "good of the species." Many nonscientists and more than a few scientists fall into a trap of thinking that if a trait furthers the survival and reproduction of other members of a species, even if it causes harm to the individual bearing it, the trait will persist. Again, a few moments of thought reveal the fallacy of this argument; a lemming flinging itself into the sea to reduce a crowded population also drowns the tendency for self-sacrifice, leaving behind its more selfish compatriots. Some ways around this problem exist, including helping one's relatives, as I mentioned earlier, but any arguments for the adaptive nature of homosexuality cannot use betterment of others as the sole justification of the trait.

Bagemihl dismisses all these arguments, at least for animals. With a starting point of aboriginal ideas about gender and sex roles, he suggests that we need to take a broader view of the role of sexuality in culture, both human and nonhuman. He notes that exclusive homosexuality has different evolutionary consequences than occasional same-sex activity, but he sidesteps biology's ultimate f-word: fitness. When evolutionary biologists talk about fitness, they do not mean hours spent at the gym, and they do not mean aptitude. They mean genes left in succeeding generations, so that a weedy male who fathers many offspring has higher fitness, broadly speaking, than a muscular one who fathers few. The genes for weediness will thus be perpetuated. Pointing out that heterosexual sex may involve violence or can be "a destructive, rather than a procreative, act," Bagemihl suggests that animals engage in homosexual activities to achieve sexual pleasure. While a reasonable proximate explanation, that is, an explanation that gives the immediate reason for performing an act at the time it occurs, this begs the question of where such activities fit into the evolutionary scheme of things. Sex is enjoyable, and different people clearly have different ideas on what constitutes enjoyable sex, but if it did not result in babies at some point we would not be doing it.

The key here is "at some point." The idea that animals, unlike humans, mate only for procreation is false. Even heterosexual sex often occurs at times when fertilization is unlikely; although this has long been understood to be the case for humans, where ovulation and hence fertility is what is termed "concealed," scientists are now realizing that other animals can show the same pattern. The lesson is that even in nonhumans, sex can be about more than reproduction. People find this surprising, and in a way it is not quite accurate, because of course ultimately everything is "about" reproduction; any trait that is not passed on will disappear. Thus foraging is about reproduction, keeping warm is about reproduction, maintaining blood pressure is about reproduction. Doing these things correctly means that the animal doing them has offspring that do them too, which is what life is all about. But even if keeping warm is about sex, none of us expect to get pregnant every time we put on a sweater. It stands to reason, then, that even sex is not always about sex, at least in the short term.

This broader viewpoint of sexual behavior—that it broadly contributes to fitness but does not have to result in offspring every time—frees us to

consider some new connections. What role does sexuality, not just reproduction, play in the lives of animals? Homosexuality might be considered
in the same light as heterosexual sex during nonfertile periods, for example.
Many forms of nonreproductive sexual behavior occur, including sex play by
juveniles, and separating homosexuality from these is counterproductive.

It also seems likely that homosexual behavior, along with other nonreproductive sex, means something different for different animals. I have
studied crickets for many years, and frequently collect individuals by listening for the distinctive courtship song that males of many species produce when they have attracted a female. This song is quieter than the
chirping that people hear on summer evenings. Mating usually follows it.
For me, the courtship song saves some work, because I can use it to localize
not one individual, as with the louder calling song, but two, and collect
both members of the pair at once.

Sometimes, however, I bend down, part the grass where the song comes
from, and discover not a male and female, but a male singing the courtship
song and another male, or a male and a juvenile cricket, either male or
female. I have always dismissed these cases as mistaken identity, part of
the slop in the system. After reading Bagemihl's book, I am not so sure. I
suppose one male courting another should be considered homosexuality.
But I do not know what it means in the lives of the crickets. They are not
social animals; aside from mating and occasional territorial chases, most
species live quite solitary lives. And every so often the courting male is
near not another cricket, even a juvenile, but a leaf or twig. Do we call
this fetishism?

I do not think we should. It makes more sense to me to conclude that
selection has acted on males to make them respond to anything vaguely
cricketlike that comes their way when they court; such behavior has generally produced more baby crickets than a trait that requires the male to
respond with greater discrimination, because the more discriminating male
risks missing a real female once in a while. However, this is not to dismiss
homosexuality the way that researchers often have. Female-female sexual
encounters in bonobos clearly play a role in social relationships among
group members. It would be ludicrous to suggest that their activities resemble a cricket singing away to a twig in the grass. The point is to see
how homosexuality, like any other behavior, fits into the lives of the organisms, not to create another category, however much that category has
been previously ignored by biased researchers. As I discussed in earlier
chapters, we do ourselves a disservice by assuming that everyone is typical,

or normal, or average. This is not so different from the sociological attitude about homosexuality and gender having many manifestations in various cultures, but it acknowledges that organisms, both humans and nonhumans, are biological entities, and any attempt to explain sexual behavior, even in its broadest sense, must be rooted in their biology.

Twelve

ONE OF THE MAJOR BATTLEGROUNDS in arguments over the existence of sex differences is mathematics ability and performance. Are boys better than girls? If so, what caused the difference? Here is an area where the strands I have been following can be seen to knot themselves almost impenetrably. What has sexual selection to do with calculus? If we are looking at a real trait here, is it adaptive, or a by-product? Should we look to evolutionary psychology for an explanation? Or are the questions being asked somehow biased? Boys and girls are different "biologically," whatever that means. They are also different in the way they experience and are experienced by society. It would be impossible to perform an experiment to determine if the difference in a particular skill were genetic or learned in boys versus girls. But how do we weight the importance of possible answers to the questions involved?

The stakes are high. Being good at math is seen as perhaps the purest indication of intellectual ability, and it is often imbued with a mystique that extends far beyond the actual skills involved. People who are good at math are popularly seen as smarter than people who are good at sculpture or auto mechanics. Even if one cannot mold a sitting figure or repair a transmission, the process of doing so is not so mysterious that it is impossible to imagine ever doing it oneself. The same is not true for mathematics, perhaps because no day-to-day analogy for the skills involved ever comes to mind; sculpture is reasonably like making sand castles, auto repair is similar to changing a light bulb or tightening a screw. Mathematicians do not spend all day adding up columns of numbers or performing long division, so what exactly do they do?

Mathematics is enigmatic, and yet it is necessary for the operation of many common objects in our lives, like computers and television sets. We say, "It's not rocket science" to mean that something is not difficult, the assumption being that rocket science requires a lot of mathematical reasoning, which by definition means it must be hard. We don't say, "It's not sonnet writing," even though most people would have at least as hard a time writing a decent sonnet as sending a space shuttle into the sky. On a practical level, quantitative skills are a requirement for many types of professions. In my own field of biology, people who use mathematical models to understand the natural world are viewed as doing "hard" science, and it is perceived as hard in both senses of the word. It is difficult, and it is closer to physics and chemistry than the so-called soft science of the rest of biology.

Like many valuable attributes, mathematical ability is commonly thought to be greater in males. Many girls and women are said to have "math anxiety," a psychological disorder or syndrome in which they are exaggeratedly fearful of quantitative tasks. Also like many valuable attributes, mathematical ability, or a counterpart to it, spatial learning, has been extensively studied in nonhuman animals, where it likewise shows a sex difference in some species. In this chapter I explore the sources of the differences in this ability between males and females, both in our own and other species, and review some of the suggested evolutionary explanations for our behavior. I also show how many of the studies purporting to explain sex differences in mathematical ability—not performance, but raw aptitude—have been flawed, partly because, as always, the environmental and genetic or physiological influences are not easily separated.

ADDING UP, TAKING AWAY

Several studies have found that girls start to perform worse than boys in mathematics at relatively early ages. Interestingly, the difference between the sexes depends on the test being administered and has shrunk overall in the last few decades, which immediately suggests that factors other than inherent skill are at play. Perhaps more important, surveys of parents show that both mothers and fathers expect that their sons will do better at math than their daughters will, regardless of evidence supporting or contradicting this assertion. These divergent expectations start when the children are very young; parents of four- and five-year-old boys predicted that their sons would solve numerical tasks more quickly than did the parents of daughters of the same age.

In 1980 Camilla Benbow and Julian Stanley published a study of mathematics performance by precocious seventh and eighth graders in the United States. They administered the Scholastic Aptitude Test (SAT) to the students, a test normally taken by students in the eleventh and twelfth grades. Because the students had not been exposed to the mathematics courses ordinarily assumed to have been part of the curriculum of the test-takers, Benbow and Stanley posited that high achievement on the test represented reasoning ability and not simply whether the students had remembered what they learned in class. The test was taken by both boys and girls, and the boys consistently scored higher than the girls in the mathematics section. Boys not only had higher average scores, their highest scores were always higher than the highest girls' scores, and they tended to show more variability in the scores they obtained as well. The authors concluded that, since educational opportunities and mathematical experience were equivalent for both sexes, these could not explain the disparity. Instead, they suggested that intrinsic differences in ability might at least contribute to the difference in performance, and although they never used the words "genetic" or "biological," this was the implication many readers received.

The work received a great deal of publicity, and was followed up in *New Scientist,* a popular science journal published in the United Kingdom, with a story under the headline "Mathematical Genius: In the Hormones?" In it, Benbow suggested, "Brilliant mathematicians are likely to be male, to suffer from allergies, to be left-handed and to be myopic." These differences were said not to be due to social influence, either. The link to hay fever comes from the discovery that "extremely precocious mathematical reasoners were about twice as likely to have allergies as members of the general population." Since other work had connected immune disorders (as well as left-handedness) to fetal exposure to testosterone, the biological connection seemed clear, though I know of no recent research that has followed up on this suggestion. Indeed, several people suggested that children with allergies or myopia might be forced to stay indoors and hence might concentrate on their schoolwork rather than playing sports, which would reverse the cause and effect in the relationship. Nevertheless, when a new version of the Barbie doll started complaining, "Math is hard," it was no more than many people had been led to expect, even though complaints about the blatant stereotyping eventually led the manufacturer, Mattell, to cancel further production of the toy.

Again, this could be viewed as an application of biology to human behavior that ends up devaluing women. According to later feminist critics of the work, some parents, at least, were sufficiently influenced by the Benbow and Stanley study that they used it to justify not encouraging their daughters to succeed in mathematics, an alarming outcome that the original authors clearly did not intend. Their conclusions were far more tempered, in part because they were studying a highly select group of students already designated as unusually gifted. What, if anything, the results of testing such a nonrepresentative sample imply for the run-of-the-mill seventh-grader is not clear. It is possible, for example, that gifted boys differ from gifted girls but that other children show no such sex difference. And Benbow and Stanley were not attempting to tell parents how to teach their very young children, or testing whether early intervention directed at minimizing the difference would work, although many people were quick to make the connection. The motivations and proscriptions of the researchers notwithstanding, numerous educators and feminist scholars objected vigorously to the work, although their rebuttals did not appear in any forum nearly as visible as the prestigious journal *Science*, which published the original study.

Girls indisputably receive a great many negative images about females and math, and teachers treat the sexes differently in the classroom. Even when girls are positive about school in general, an Australian study showed that they were still negative about math and science. Separating the sexes in the classroom did not seem to help, perhaps because the images of mathematics are so counter to images of femininity that they cannot be overcome in the hour a day students spend learning to solve equations. Furthermore, although neither the boys nor the girls in the study made a connection between taking courses in math and their career goals, the girls were less likely to end up taking those courses in the first place. Girls also sometimes have lower expectations for their own performance in math, and when they do poorly, they attribute their failure to a lack of ability, not to external factors like an unfair test, an unreasonable or poor teacher, or simple luck. Boys, on the other hand, tend to attribute success to their own skill but failure to irrational forces like the teacher having a bad day when the test was given.

A few educational programs have examined the way in which different students learn mathematical skills, and suggested that boys and girls may require being taught the same material in different ways to understand it equally. It is often suggested that girls learn more in a "cooperative" atmosphere where competition for the right answer is downplayed, and teachers are exhorted to make mathematics more relevant to everyday life. Some of the new computer games purport to attract girls by emphasizing relationships rather than alien-bashing. This notion hints at intrinsic differences between the sexes, though it does not give any clues about where such differences might have come from. I am leery of this approach for some of the same reasons I am suspicious of ecofeminism: do we need to invoke girl-math and boy-math, even if we were to strive to make our attitudes more egalitarian? I remember clearly how a chemistry teacher in seventh grade tried to tell me that solving a certain problem was "just like baking a cake" in an attempt (I assume) to reassure me of my ability to master it. Even at twelve I found the assumption that I would relate to cooking with familiarity to be questionable, never, as it happened, having baked a cake in my life.

Nowadays, the mathematics education literature contains copious amounts of advice and information on increasing the numbers of girls taking math, improving their attitudes toward it, and discovering the impediments to their learning it, though none of it even hints at a potential for a biological explanation for a gender gap in math achievement scores. This seems to me, once again, to be getting to a right, or at least socially satisfying, answer for the wrong reasons. Yes, we need to find out the impediments to girls learning quantitative skills, but might we also need to look squarely at the potential for a biological explanation? There is a silence about the possibility of inherent differences between boys and girls here that is not altogether healthy; the idea lurks in the background. I suspect that researchers shy away from biological explanations, any biological explanations, simply because the one that was proffered led to an answer no one wanted to hear. I personally did not want to hear it either, but I think the answer is flawed for scientific reasons, not political ones. I think we need to confront the question of whether boys—or men—are naturally better at mathematics on its own terms, by examining the kind of answer we can expect to this type of question. Rather than shying away from such studies, we need to examine their assumptions more closely. Can we ever answer a question about "inherent" ability? What would we need to know to do so?

First of all, remember the distinction I made in an earlier chapter about whether a trait itself or a difference between traits can be said to be genetic or learned. The difference between traits can be declared one or the other if a test can hold everything else constant, as in the hypothetical example of the identical twins being reared in different countries and coming to speak different languages. In the present case, a difference between boys and girls could be said to be inherited if the only difference between the two groups was their genetic makeup, with all environmental effects the same. Such a situation is both practically and theoretically impossible, however, which is one of the things that makes the supposed nature-nurture debate so silly; boys and girls are treated differently from the moment they are born, and even before birth may be subject to differing influences from maternal genes. Although Benbow and Stanley attempted to control for such variable environments by ensuring that the middle school students they studied had experienced similar mathematics courses and had the same expressed fondness (or lack thereof) for numbers, it is clearly a stretch to suggest that these traits constitute all the social factors influencing one's ability to perform on an achievement test. An alternative explanation for their findings is that the different experiences of boys and girls led them to attain different scores.

Interestingly, although we lack the controlled experiments of boys and girls given identical environments, we do have information on sex differences in mathematical achievement and ability across cultures. Achievement, as I mentioned, refers to mastery of material learned in school, while ability attempts to measure how well people can use their knowledge about mathematics to solve new problems. Benbow and Stanley were concerned with ability; tests of mathematical achievement tend to show smaller or no sex differences. Children from the United States routinely do worse on achievement tests than children from Japan and Taiwan, and on tests of ability they also score lower than children from several other countries. Sex differences appear in other cultures besides the United States, and they are always in the same direction, with girls being worse at solving certain types of problems. A crucial point, however, is that the sex differences are always smaller than the differences across cultures, so that Japanese girls, for example, outperform American boys of the same age by a wider margin than the American boys outperform American girls. Such measurement of variation is important, and I will return to it later.

What, then, do we do with the results of the various tests? It is short-sighted to dismiss any biologically based explanation as chauvinistic clap-trap simply because it is biological. The worry, of course, is that if we believe any part of a trait is genetic, we will be less inclined to use social means to change it. Yet the evidence suggests that it does indeed make sense to intervene in the ways that we can, perhaps by making mathematics more accessible to girls along the lines suggested by the mathematics educators. Declaring that such interventions are hopeless because math ability is not learned is not just sexist, it is not supported by the data. Rejecting the biological explanations merely because they are biological is to give them more power than they deserve.

FINDING THE WAY

Although the search for biological differences between the sexes in mathematical ability has been largely abandoned by the mathematics educators, it has taken another form, and one in which evolutionary biology has played an important role. One of the reasons boys and men reputedly have greater mathematical ability is that they are supposed to be better at spatial learning, or visualizing problems in multiple dimensions. On the mathematical ability tests administered in many of the studies I mentioned previously, the problems that girls tend to do worse at are those involving, for example, mental rotation of a drawing of a three-dimensional object. Sometimes the sex difference is generalized to the use of maps, or the ability to navigate in a car or while flying an airplane.

Numerous studies have attempted to measure the magnitude and nature of this difference between males and females, and a few have purported to explain it in evolutionary terms, so that our history as hunters and gatherers or as a polygynous species accounts for males' greater skill at visualizing which of an array of drawings is most like a model once it has been rotated. One of the best attempts to synthesize the research which assessed spatial ability was a meta-analysis performed by D. Voyer and coauthors. A meta-analysis takes the results of a compendium of tests of the same phenomenon and examines the entire data set for statistically meaningful conclusions. It is essentially a test of other tests which, because it incorporates the results of many different types of work, can be very powerful. Variations in, for example, the statistical methods used to evaluate each test, or the sample sizes, or the laboratory setting in which the tests were administered, become unimportant when a large enough group of studies is combined. Voyer et al. looked at 286 different reports of spatial ability published up to 1993, which compared males and females. The data contained several types of tests, including the aforementioned mental rotation, a paper folding task, and the "rod-and-frame" test, which requires adjusting a rod to a vertical position inside a tilted frame (the frame therefore provides no clue to the location of the vertical plane).

The researchers found that some tests consistently showed sex differences, while others did not. Where differences existed, boys or men always outperformed girls and women, although some tests showed a difference at some ages and not others. As with mathematical ability, the difference between the sexes appeared to be narrowing, at least in some cases, so that tests administered twenty years ago were more likely to find larger differences than tests administered recently. They concluded that the difference was real, but made no attempt to explain it, that not being their goal.

Other work has examined the effect of sex hormones on spatial learning, although as I describe below, much of the connection between brain physiology and performance comes from experiments on nonhuman species. Again, although the studies are not always identical in their findings, a consensus appears to be emerging that spatial ability has some links to structures and chemicals in the brain, and that those structures and chemicals differ in a few ways between the sexes.

What lesson should we take from these studies, and how do they apply to ideas about the likelihood that women will succeed in aerospace engineering? Does a sex difference in mental rotation ability mean anything about our future, or is it an unimportant blip in the litany of traits making us all unique? One reason why such differences have been suggested to dog our present-day activities is their supposed connection to our prehistoric ancestry. If early humans showed a sex difference that could be demonstrated to be adaptive, at least some scientists then imply that we are stuck with it, long after the need to bring down a mammoth has faded. Many biologists assume that if a trait is persistent and appears to have utility in some environments, it must be an adaptation, although as I discussed in the chapter on female orgasm, such an assumption has also been questioned. Is the male ability to do mental rotations an adaptation left over from our evolutionary past? If so, what does it mean for our expectations of female performance? Well before the Voyer et al. analysis was published and in several papers since, numerous psychologists and anthropologists were taking the difference between the sexes in their performance on the tests as a given, speculating on its significance to daily life, and finding evolutionary explanations for it.

A 1971 paper by the psychologists Jeffrey Gray and Anthony Buffery claimed that several generalizations about sex differences could be explained by examining the adaptive advantage of possessing each differentiated skill. They examined four findings that they said had been substantiated in previous studies: greater aggressiveness in male mammals; greater fearfulness in male rodents but also in female humans and "perhaps in primates generally"; mammalian male superiority at spatial tasks; and greater female linguistic ability in humans (or "man," as they put it).

The robustness of all these generalizations is questionable, with, as we have seen, the possible exception of the third. Males are more aggressive, Gray and Buffery claim, because male animals always have dominance hierarchies into which females are subsumed and in which females are always subordinate. As it turns out, this is quite untrue, as I discussed in the chapter on dominance, and furthermore the significance of dominance hierarchies is uncertain for many natural populations of animals. The psychologists seem to be suffering from misconceptions about a *scala naturae* as well; they conclude that a phenomenon seen in rats, mice, a monkey or two, and some subgroups of humans must perforce be general to all mammals.

The reasoning behind the claim that female rats and mice are less fearful than males of the same species and more fearful than human males is a bit hard to follow, but seems to have to do with women being "more prone than men to phobias (and especially agoraphobia) and reactive depression; they also score more highly than men on personality tests measuring neuroticism and introversion . . . as well as on tests measuring . . . susceptibility to anxiety" (p. 98). The evolutionary justification for this is unclear; is being fearful or being bold the adaptation? The answer would seem to depend on the environmental circumstances. In addition, the authors do not seem to consider that social influences and discrimination against women might enter into some of the findings. Many feminist scholars, such as Phyllis Chesler, have argued that depression, for example, is an expected and in many ways reasonable response to an unjust world. Regardless, biologists have not followed up this particular sexual dimorphism, either in animals or in humans.

Gray and Buffery thought that the greater linguistic ability of women stemmed from the prolonged mother-infant bond that occurs in humans; they suggested that mothers were thus in a better position to teach their children to speak than any man would be, and hence women were subject to differential selection because they were the instructors for the society. Why such a sophisticated ability would evolve only to be squandered on individuals babbling in words of single syllables is again not addressed. Neither is the problem of teaching sons, boys presumably being the less adept sex at learning the verbal skills of their mothers. Furthermore, like male dominance, female superiority in verbal ability has not been consistently demonstrated, and many educators and psychologists feel that little if any sex difference in this trait actually exists. Perhaps most telling, even when many people think that girls develop greater verbal ability than boys, no one seems to find it incongruous that in the humanities, as in the sciences, males dominate the top professional ranks.

We are therefore left with the sex difference in spatial learning, which alone has hung on as separating the girls from the boys. Here the rationale, voiced by Gray and Buffery but echoed in other papers, is that males have larger territories or home ranges than females because they move among the areas occupied by several females in an effort to find mates. Selection therefore favors a greater ability to navigate in males. This idea lends itself to testing in nonhumans, since it suggests that the sex difference should not occur in species lacking such a disparity in male and female home ranges.

To confine our attention to humans for the moment, however, additional adaptive explanations for the greater spatial ability of males have been proposed. David Sherry and Elizabeth Hampson, psychologists from the University of Western Ontario, summarized seven of these. As well as the home range size difference, they list a "male foraging" hypothesis, which suggests that men needed more skill at navigation and map-reading because it helped in the hunt, and a "female foraging" hypothesis, which actually predicts greater spatial abilities in women, at least in remembering where arrays of objects are located, because of the need for finding non-mobile food. Sherry and Hampson note that a few tests have indeed shown that women excel at such tasks, but the overall greater interest in and information about the difference between the sexes at more three-dimensional tests overshadowed this area of inquiry.

Sherry and Hampson also mention the "male warfare" and "female choice" hypotheses; according to the former, males benefit by having larger ranges, but for the purpose of attacking other men, not because they are cruising for females. The latter suggests that females prefer good hunters, and spatial ability again aids in hunting. Males capable of hunting for game would have been viewed as desirable husbands, with their skill serving as a kind of display, like the courtship dances of a male sage grouse, indicating their prowess. Another hypothesis, the "dispersal" hypothesis, is similar to the female foraging idea, because it too predicts that women will be better at spatial learning than men, in this case because they are the sex that disperses farther from the area where they were born and hence benefit from sophisticated orientation skills. Again, since the overall evidence contradicts the prediction arising from this hypothesis, Sherry and Hampson do not dwell on this idea.

Finally, the two put forth their own explanation, the "fertility and parental care" hypothesis. According to this, women are better off having worse spatial skills because they are therefore more likely to stay close to home, particularly during pregnancy and lactation. This reduced mobility in turn increases their safety and allows them to save energy for reproduction.

Sherry and Hampson compare the predictions of the various hypotheses in areas such as the likelihood that the sex difference in spatial ability will change at puberty or with menopause and old age. They find, not surprisingly, that their own hypothesis is best supported by the data; the hormonal changes associated with performance on spatial skills tests are indeed those that change at puberty, when a woman becomes capable of reproduction. Less evidence is available on the changes in spatial learning ability that may occur late in life, perhaps because it is more difficult to obtain willing senior citizens than elementary and college students for such experiments.

I am more than a little skeptical about all these explanations, and especially about the focus on their significance in our day-to-day lives. First, has anyone ever demonstrated that the small differences in mental rotation skills are valuable, let alone that they would be adaptive and could even hypothetically increase reproductive success in the sex with the higher test score? In modern society we tend to favor high grades for their own sake and for what they symbolize on college and job applications, but more than symbolism is needed for a trait to be a demonstrated adaptation. Second, why wouldn't any advantage for men also be useful, perhaps in a different context, for women? Is maintaining spatial ability in the brain costly, so that it would be lost when not absolutely necessary? No one knows, though the data on other species suggest that it might be, at least if the difference between the sexes is large. Sherry and Hampson's hypothesis does focus on a disadvantage to females, rather than an advantage to males, but it seems to me that women might stay put during pregnancy because they weigh up to 25 percent more than they did before they conceived, not because they are more likely to get lost in the woods. In any event, reduced wanderlust is a separate characteristic from lack of navigational skill; not wanting to do something is not always the outcome of being bad at it. Third, the hypotheses that deal with foraging seem to assume that all hunters hunt alike, when in fact contemporary hunter-gatherer societies vary widely in the extent to which men range over large areas and find game essential to the survival of their group. The hormonal associations with spatial ability may simply be a holdover, an epiphenomenon significant in some species and contexts but not others.

I am not denying the existence of the sex difference, just suggesting that the explanations are post hoc, and do not necessarily enhance our understanding of male or female biology. The problem is that such explanations permeate our society, and it is important to recognize that we can examine the conclusions and use an awareness of our biases to evaluate them. On the one hand, we should not ban research on sex differences by claiming that it inevitably harms women. On the other, it is foolish to uncritically accept the assumptions—that early female humans did not wander, that men were dominant over women—as the only ones possible for constructing evolutionary hypotheses.

OF VOLES AND MEN

Although animals do not use paper maps or take SAT tests, similar skills are involved in negotiating mazes, and tests of animals' ability to find a reward or an exit in a maze have been used to determine the spatial learning skills of a variety of species, particularly rodents. A particularly popular form of this learning test uses a Morris water maze, in which animals learn to find an underwater platform by swimming to it in a circular maze filled with water made opaque through the addition of latex or other inert material. The fewer trials needed to find the platform, the greater the spatial ability. In some species, there is again a sex difference, with males performing better than females, at least in rats and some voles, a group of rodents which includes the lemmings and which has representatives in many parts of the world. Among the voles, however, species vary in whether they exhibit the difference or not.

The meadow vole, *Microtus pennsylvanicus,* is polygynous; a male will mate with more than one female as he travels over an area that includes the ranges of several of them. In contrast, the prairie and pine voles, *M. ochrogaster* and *M. pinetorum,* are both monogamous, and males and females stay together in a territory. When the males and females of all three species were tested for their ability to learn mazes, the monogamous species showed no sex difference. The male meadow voles, however, were able to learn their way through seven different kinds of mazes more quickly than the females, an ability that one expects to be advantageous given the greater roaming shown by the polygynous males. Females with low estradiol (the rodent equivalent of estrogen) levels did better than females with high levels of the hormone, and no sex differences were apparent in juveniles. When deer mice, *Peromyscus maniculatus,* another species of rodent found in many parts of the New World, were tested in the Morris water maze, females did better outside the breeding season and worse during it, while males showed the reverse of this pattern. In accordance with the difference in learning, the hippocampus, the part of the brain used in spatial skills, is larger in male meadow voles than it is in females, while the monogamous species show no such dimorphism in hippocampus size. Here we can actually compare a difference in a trait, and not the trait itself, and the biological basis is quite clear.

These links between time of year, brain structure, and hormone levels have made the situation more complicated in studies of the favorite model organism of experimental psychologists, the lab rat. Although some tests of laboratory rats had shown that males were better at spatial tasks than females, recent work examining females over the course of their estrous cycle revealed that female performance varied depending on their reproductive state, whereas males were relatively constant in ability. Overall, no significant difference between the sexes could be detected, but had the investigators studied rats across several cycles and then compared males and females at each day, they would have either found no difference or, on certain days, slightly better performance by the males. Many scientists would have concluded that the difference was more meaningful than the similarity, although the authors of this study, by explicitly taking estrus status into account, were able to conclude that the sex difference was slight, and vanished when females were averaged across their cycle. Females were not less active when they were receptive, which contradicts the fertility and parental care hypothesis of Sherry and Hampson.

Note that there is nothing fixed about the association between being male and doing well at spatial learning or having a larger hippocampus. Females sometimes use more complex navigational skills than males. In the brown-headed cowbird *(Molothrus ater),* a brood parasite which lays eggs in the nests of other species, females need to search the environment for potential hosts. They may come back to nests more than once, because the development of their eggs needs to be synchronized with that of the host eggs to ensure that the host will see cowbird chicks alongside its own chicks and not alongside eggs, which trigger a quite different response.

The cowbirds therefore need to remember where each host nest was and be able to return to it at the appropriate time. All of this requires quite sophisticated spatial learning, and therefore it is no surprise that female cowbirds have a larger hippocampus than male cowbirds. The hormonal mechanisms underlying these behaviors are not as well understood as those in the rodents, but it is worth remembering that birds also have testosterone and estradiol as the major sex hormones underlying reproductive behavior and physiology. Testosterone clearly does not dictate male superiority at spatial tasks; it is just that selection is likely to make use of whatever mechanisms are available, and hormonal differences between the sexes are widespread among vertebrates.

SEXUAL NAVIGATION LESSONS

The lesson from the work on birds and rodents is that the life history of a species can often make sense of sex differences, as I discussed earlier with regard to qualities such as dominance and maternal care. But how is the difference between male and female voles, deer mice, and cowbirds relevant to girls taking fewer math classes than boys, or to the likelihood that men and women engineers will ever be equally common?

Men are far more like women than they are like voles, even male voles. This statement seems to require much more vigorous assertion than one might think, perhaps because we are often so intent on scrutinizing the world for differences that support our biases. Our hormones change with the menstrual cycle, but they vary far less in magnitude than the same hormones in any seasonally reproducing animal. The differences between a breeding and nonbreeding white-crowned sparrow or vole are profound. The differences between an ovulating woman and a menstruating one are minuscule in comparison. Not nonexistent, just tiny. Similarly, hormones appear to influence many aspects of human behavior, including spatial learning ability, but the sexes differ far less than is expected in a more seasonal organism. I am not convinced that in trying to find an adaptive advantage to these sex differences in humans we are not chasing after artifacts of our mammalian heritage that have little bearing on our lives both then and now. We would need to devise a testable prediction based on the hypothesis that greater male spatial ability yielded higher reproductive success; although ways to do so without going back in time exist, such predictions have not yet been made.

Perhaps it was historically advantageous for men to find their way over longer distances than women, perhaps not. In any case, it is important to remember two things. First, the degree of the difference between men and women or boys and girls on either spatial learning tasks or other tests of mathematical ability is quite small. In every case, performance by the sexes is largely overlapping, so that one cannot predict the sex of an individual if one knows only what score that individual received on a test. This is not the same as saying there is no difference; Benbow and Stanley indubitably obtained one in their test. How meaningful the differences are is another story. Take height, for example, a trait on which men and women obviously differ. Nevertheless, knowing only the height of an individual and nothing else about the person—not age, not ethnicity, not nutritional status during development—would give us a very poor predictor of sex. The narrowing gap between boys' and girls' performance on many mathematics achievement tests over only a few decades also argues against a very strong role for biology in explaining much about the huge difference in, for example, the number of men and women physicists. Social influences are likely to be more important here. We should not be seduced by the plausibility of evolutionary explanations for differences between male and female voles in their spatial learning into automatically generalizing to men and women, as I have been cautioning throughout this book.

The second sticking point at accepting the difference between men and women in mental rotation tasks as meaningful in our lives is that no one, including the many scientists vigorously studying the problem, has ever demonstrated the existence of a connection between performance on these tests and anything that happens later in life, including choosing a career, getting promoted to vice president or associate professor, or being able to understand a map of how to get to New Jersey. The scientists are not studying the problem so that they can justify parents encouraging their sons more than their daughters in school, and we should not be fearful of their findings. They are studying it because understanding what differences exist between groups is interesting, and because if we document the differences we can have some hope of understanding their basis. Though we need to ask who has a stake in the measurements as well as to look at the measurements themselves, we do not need more fodder for the gender war cannons, whether taken from other animals or from our own children. We can find differences without attaching value to them, and without exaggerating their importance in our daily lives. This means keeping an open mind and realizing that our preconceptions can make us ignore some data while accepting others. Many people still believe that girls are better at verbal skills than boys, but they do not voice amazement at Shakespeare being a man or suggest that he must have had a female ghost writer. This should provide a clue that, whether the sexes differ in verbal ability or not, what we see is the result of bias, not biology.

Conclusion

UNNATURAL BOUNDARIES

I HAVE SPENT MOST OF THIS BOOK—and a reasonable part of my career —speaking simultaneously to feminists and scientists, in the hope that the connection between the two can become more illuminating and less fractious. I will conclude with the same hope, first voiced by Patty Gowaty, that "Darwinian feminism is an oxymoron no longer." Steering a course between two sometimes hostile entities is fraught with risk, but the opportunities for cross-fertilization, to use yet another biologically sexual metaphor, are too great to pass up. I have tried in the previous chapters to see how views of gender in humans color our ability to understand nature, and similarly how our views of animals affect how we see ourselves. Because we obtain ideas on gender in animals in large part from our ideas about humans, this is a tricky task, and it is made even more complicated by our being animals, subject to the same selective pressures that mold the behavior of other creatures. As is clear from the preceding pages, I accept modern science as a discipline worth pursuing, and fall into the camp of what some would refer to as liberal feminism, in that I believe that our gender biases have not hopelessly interfered with our ability to understand nature.

How, then, do we use animal behavior to teach us about ourselves? In its simplest form, we see what animals do, whether it is to remain mated to one individual for life, to commit infanticide, or to cooperate in the rearing of young, and say to ourselves, "Humans do that—or should do that, or shouldn't do that—too." The question is what to do with this recognition. Many feminist critics have decried simplistic storytelling in which scientists, at least according to their analysis, simply assumed that humans did things for the same reasons that animals did, and worse yet, that because the behaviors occurred "naturally" in the animals, we could safely extrapolate and conclude that the human equivalents occurred for the same reasons and were equally easy to justify. Hence the popular notions that males want to play the field and women seek monogamy, or that sexual violence is an inevitable part of our lives. After all, it happens that way in ducks, or hamsters, or fruit flies. The media exploit any tendency for scientists to draw these kinds of analogies, because headlines like "Rape is natural" or "Fooling around is in your genes" attract more attention than the more accurate "Some species respond to certain selective pressures by increased levels of extra-pair copulations at the time of ovulation." Certainly some scientists have not bothered to correct these oversimplifications and inaccuracies, and others seem to willingly compromise their standards; I heard one biologist author of a best-selling book on sex declare at an international scientific congress that he simply did not think the public could comprehend the subtleties of the scientific research, so it did not matter if we gave them incorrect or sensationalized information.

Because of such sloppiness, it is no surprise that many people get twitchy when they hear about any so-called biological explanations of behavior in humans and are also nervous when scientists study behaviors in animals that seem to occur in humans. In this book I have tried to take away this fear, though at least in part it should be replaced with caution and awareness. Biology has great potential for harming women, but that is owing to human misuse of the science, not the science itself. Here I examine some of the ways in which discoveries about animal behavior have been misapplied in discussions of gender by both parties in the battle of the sexes. I suggest that feminism has more to offer biology than biology does feminism, but that biology nonetheless has a large role to play in helping us to understand human—as well as animal—sex differences and similarities. Together, feminism and biology can extend the boundaries of our thinking about gender in ways that neither could accomplish alone.

SOMETIMES A SNAKE IS JUST A SNAKE

In our zeal to make animal behavior interesting and relevant to the public, scientists often summarize research with aphorisms about its meaning in the context of human life. Because the media are eager to capitalize on this tendency, studies of mixed paternity in blackbirds get a spin into human adultery. As a result, discoveries about animal behavior are often expected to "mean" something, much as animals are expected to "do" something. Thus learning that female adders mate with many males, and indeed that some species of snakes form writhing balls of mating individuals, must imply that—what? Orgies are natural? Sexually voracious females are to be applauded?

There is not a moral to every story in animal behavior. Sometimes a snake is just a snake, and sometimes snake sex is only about sex in snakes, or sex in egg-laying reptiles. Although a biologist's job in part is to interpret what organisms do in a broader context, that context does not, and should not, need to include a lesson for human beings. This is true regardless of whether the lesson is something we would like to teach, which means that using animals as vehicles for nonsexist thinking is just as out of bounds as using them to keep women barefoot and pregnant. Our attachment to the *scala naturae* makes us more likely to use certain animals as object lessons, which compounds the problem; male dominance in baboons or chimpanzees seems so much more compelling than male submission in spiders or fishes. In certain contexts, primates *are* more relevant than pipefishes. It would be foolish to study cognition in a slug. But slugs can show us that it is not necessary to have cognition to exhibit, say, parental care. Feminists do not have to buy into the idea that animal behavior is more ammunition in a war for or against equality of the sexes. I have been arguing throughout this book that we lose in several ways by exploiting animals' activities, whether for or against apparent egalitarianism. This is what I mean by biology having less to offer feminism than vice versa; discoveries of mate fidelity, male tenderness, or female sexual violence do not argue for a human nature that includes or excludes them. Instead, feminism can suggest that our perception of how animals behave is colored by our perceptions of our own behavior, a suggestion that can help structure our science. That scientific information is misused is undeniable, but our response must not be to throw out the baby with the biased bathwater.

FIRST IMPRESSIONS

An example of how information about animals has been used in detrimental ways to women when it did not have to be comes from work, old and new, on how mothers and offspring interact immediately after birth or hatching. Most invertebrate offspring grow up alone, of course, and the same is true for many vertebrates. But in a diversity of species, a process called "imprinting" occurs. This has different manifestations in different animals, and what you see also depends on whether you take the viewpoint of the offspring or the mother. The phenomenon, originally studied in ducks and geese, was made famous by the Nobel laureate Konrad Lorenz, who was photographed leading a line of downy little goslings to the water to swim. The idea is that the goslings, or ducklings, or domestic chicks, will follow an object that they see during a crucial period early in life. Back in the 1930s Lorenz originally specified two aspects to this process: the narrow window of time during which the young animal forms the attachment, after which no imprinting stimulus is effective; and the irreversible nature of the attachment—once a young animal imprints on something, it is permanent and cannot be altered. The goslings following the scientist had been exposed to Lorenz instead of their mother, and hence imprinted on him and would have stared uncomprehendingly at any nearby goose soliciting their attention.

Since his time other scientists have demonstrated some leeway to these general principles, such as the influence of sounds heard before hatching on later following responses, but they are still broadly applicable. The period during which an attachment can be formed is called the sensitive period, and the portion of it during which the performance and reinforcement of the attachment response is the greatest is called the critical period. These times are species-specific. For example, in mallard ducks, the sensitive period is from about five to twenty-four hours after hatching, and the critical period, when imprinting is most successful, occurs between fourteen and sixteen hours after hatching.

What happens during imprinting? For ducks and geese the formula is simple: follow the object you imprint on. In nature, of course, this object is the mother duck or goose, but in experiments chicks can imprint upon other animals, say a bird of another species, a famous scientist, or even an inanimate object like a ball or a toy car. Some models are better than others at eliciting the response with a given amount of exposure. Again, this differs for different species, and in many birds—such as songbirds reared in a nest, as opposed to birds that walk around with their parent— imprinting in the classic sense is completely absent.

A slightly different take on the same idea is used by farmers to encourage the "adoption" of orphaned animals. Under ordinary circumstances, many ungulate mothers lick their babies immediately after they emerge from the womb; this process allows the mother to imprint on the smell of her offspring. She will not suckle a lamb or calf that does not have that distinctive odor. If, however, a lamb has lost its mother, as long as you cover it with amniotic fluid from a delivering ewe, you can show the lamb to a ewe shortly after she has given birth and they will both act as if the lamb is hers. If you try after the sensitive period is over, they will not. Again, this behavior is seen in some species and not others, and it never occurs in many other mammals, particularly those with young more helpless than the relatively precocious lambs and goats.

Imprinting, a classic concept in animal behavior that illustrates how learning interacts with the environment from the very start of life, becomes relevant to my point in this chapter because of the way ideas about it have been applied to humans. For reasons that are not entirely clear, a few decades ago imprinting became connected in people's minds with attachment theory, a body of work in psychology dealing with the way in which mothers and newborns develop a close relationship.

In her masterful book *Mother Nature*, Sarah Blaffer Hrdy discusses the significance of attachment, originally promoted by the psychologist John Bowlby, and its relationship to current ideas about motherhood, "bonding," and by extension to topics such as day care and the likelihood of child abuse by parents and other caregivers. Attachment theory was directed at understanding how the baby learns to fixate on a single individual, generally the mother, at a very early age. Bonding came to be viewed similarly, but oddly enough interest in it was concentrated on the mother learning to focus on her child to maximize the likelihood of a strong connection between them. Advice to mothers began to include suggestions about physical contact with infants immediately after birth, in contrast to the earlier Western practice of separating babies from their mothers and placing them in hospital nurseries, and, as is frequently the case, advice turned into dogma. Hrdy says, "What had been presented to me at the birth of my first baby as a welcome option, had by the births of numbers two and three been elevated to 'advisable'. At the extreme end of this movement, some new mothers (whether they felt like it or not) were instructed to engage in set amounts of flesh-against-flesh intimacy, beginning right after birth, to ensure that 'bonding' took place. . . . By the 1990s, the pendulum had swung so far that in some hospitals new mothers were being rated: 'bonded' or 'not bonded'" (p. 486). If a ewe rejected her young or goslings failed to follow their mother, the reasoning seemed to go, it was because of ineffective imprinting, so therefore maternal-infant problems might well be due to the same thing, with a slightly different title.

Information about geese and sheep was thus applied to humans in exactly the way the feminist critics of biological explanations for human behavior always feared. Its application made some new mothers feel guilty for not instantly, magically, transforming into devoted parents with an unbreakable and intuitive understanding of their child. It also may have dissuaded prospective adoptive parents, since, like Lorenz's cheeping entourage, it suggests that a relationship not established during the critical time window is unlikely to be deep or satisfying. Finally, it suggests that fathers will inevitably lack a close connection with their children, because males do not bond. These are all clearly undesirable outcomes, in no small part because it now appears that bonding as a concept is wildly overstated. Bonding in primates is not like imprinting in sheep. Available research does not suggest that physical contact with a baby right after birth has long-lasting effects except when the mother is already in a situation inclining her toward neglecting her child.

Such a state of affairs is exactly why many feminists abandon biology as having no relevance whatsoever to human lives. Forcing women into a role that makes them feel inadequate, or limiting their options because biology supposedly dictates it, is a long-standing and sorry legacy of science, or at least of people writing and talking about science. It would be simple and safe to conclude that because of this risk, we should stop generalizing from other animals to humans, period. Leave the geese to their goslings, the ewes to their lambs, and let mothers decide how to interact with their children.

Except the problem is not that we looked at animal behavior; the problem is that we did not look at it long or hard enough, or perhaps that despite our scrutiny we could not escape our biases. What is the evolutionary significance of imprinting in the animals in which it was originally discovered? A clue probably lies in those species that do *not* exhibit it, including most songbirds and many species of mammals with extremely helpless young. All of these non-imprinting species have young that are what is termed "altricial," as opposed to the "precocial" young of the imprinting species. Altricial young are born or hatch in a relatively helpless state; they have little or no feathers or fur, they often cannot see or hear, and they rely completely on their parents to feed them and keep them warm until they are able to move and sense the world on their own. Examples of altricial birds include common songbirds like robins and warblers; the parents bring worms to the nest because the chicks are physically incapable of getting their own. Altricial mammals include young rats, puppies, and kittens, which cannot follow their mothers to obtain milk or shelter from the elements until they have developed for some time after birth.

In contrast, precocial babies are mobile and aware of their surroundings. Within hours of hatching, ducklings can follow their mother and be shown food, which they are then capable of picking up from the ground or water on their own. Young lambs and other ungulates are equally impressive, and can walk, often run, within hours of their birth. The two classes are not completely distinct, so categories such as "semi-altricial" have been created for young animals that acquire independence relatively quickly but not with the immediacy of a gosling, but they are workable extremes of a continuum. By and large, young animals that we tend to think of as cute and cuddly right after birth are precocial and those that seem naked, ungainly, and ugly are altricial; ducklings and colts are cute and baby bluebirds and rats are not.

The interesting exception to all of this is humans. Human infants are a complex mosaic of attributes from both categories. In many ways they are unbelievably altricial; they cannot even cling to their hairless mothers except right after birth, much less wander around the kitchen and open their own cereal boxes. And yet, of course, we find them irresistible, a phenomenon that does not require much thought to justify from an evolutionary point of view, since not taking care of your offspring is a good way to ensure that your genes are not passed on. The characteristics that make us perceive certain beings as appealing are the subject of much speculation in psychology. Perhaps it is this superficial but misleading resemblance to the lambs rather than the mice, the goslings rather than the robins, that paved the way for the confusion about imprinting.

Imprinting on a particular object to follow, and rejecting any offspring other than that which one has imprinted upon, are both behaviors that are valuable in some contexts and useless in others. For a tiny duckling waddling toward the pond, it is a very bad idea to lose track of mother duck and start wandering after a passing bicyclist. Following one's mother, without distraction, without having to distinguish her from other stimuli in a new environment, is valuable. It is adaptive. Ducklings doing so are blessed with survival and reproduction above all other ducklings. Yet a similar pattern of behavior in a robin chick would be pointless; why distinguish one adult bearing food from another? Quite the reverse of the duck's situation exists; any robin chick indiscriminately gobbling worms from anyone offering them would do better than a more shy and picky nestmate. Therefore selection for imprinting is expected to occur in precocial young rather than altricial ones.

Similarly, a mother sheep in a flock has milk sufficient for her own lambs and does not benefit by feeding unrelated interlopers. Lambs can physically wander to ewes other than their mothers, and unrelated females rejecting such advances will fare better in terms of reproductive success than mothers unable to tell their lamb from a stranger. Again, however, these situations are not likely to arise when the hairless altricial rat pups seek milk, and such mothers are likely to be less discriminating. Exceptions certainly occur, such as with the dwarf mongooses nursing their nieces and nephews, and maternal recognition is often adaptive in other contexts, but by and large whether females irreversibly connect to their young within minutes or hours of birth is critically important only when a risk of mistaken identity exists.

The likelihood of a woman setting down her newborn infant and mistakenly picking up another to nurse in its stead is virtually zero. It is also virtually zero in most other primate species, as Hrdy details in her book. The bonding brigade was misled, perhaps because we identify a little more easily with what we see as cute offspring. In addition, we do not have litters of many young at a time, which also makes it harder to see ourselves as more similar to rats than to sheep.

Incidentally, the type of imprinting discussed above, attachment between parents and offspring, is called filial imprinting. Another type of imprinting, called sexual imprinting, refers to the development of an image of the appropriate mate. Mallards, for example, are reared by their mother, who differs in appearance from their father in having mottled brown plumage with a blue wing patch called a speculum, rather than the male iridescent green and gray feathers. While a chick, a male mallard will imprint sexually on the female mallard who is his mother. When he is an adult, he will direct sexual behavior toward animals with the same general appearance, which under normal circumstances is a good idea. He does not direct sexual behavior toward adult males. If an experimenter gives him a red balloon or a human during the sensitive period, he will grow up and try to mate with a red balloon or a human, which is not such a good idea.

Interestingly, female mallards do not need the stimulus of an adult male at a young age to respond appropriately to male mallards as adults. This makes sense in the context of the birds' natural history, because male mallards do not usually hang around their offspring. If a female mallard chick needed a male to look at before she could mate with the right species, she would be in bad shape, because often none is available. Other kinds of animals show this kind of sexual imprinting to greater or lesser degrees, and it has been implicated in some important evolutionary processes such as species formation.

BEYOND STEREOTYPES

The moral of the imprinting story as applied to humans is not, however, that it is always ill advised to examine animal behavior in the context of our own, or that doing so inevitably results in damage to women. Nor would it have worked out better had we taken ideas about bonding from a species like house mice that freely allows communal nursing. Perhaps those responsible for perpetrating the bonding idea had an unconscious agenda. They may have found it a useful cautionary tale to inform women that they must remain near or risk irreparable harm to their babies. Hrdy discusses the problem of guilt in mothers, ancient and modern, at length. Had researchers observed mongooses or mice and concluded that women should pass their babies around to other lactating women for the best child development, however, it would still have ended in tears, as the saying goes. Why choose one model over another?

The point is that all species evolved in a particular environment, and the adaptations they exhibit are interesting and potentially relevant to ourselves, but we need to select our models wisely and keep context in mind. It is interesting that altricial species tend to do one thing and precocial species another, but it does not follow that humans should behave like either one. What does it mean to extrapolate? Critics of biological approaches to behavior often dismiss attempts to suggest that humans act like animals by pointing out that humans are not exactly like other animals. They often invoke the *scala naturae* to dismiss parallels with species other than primates. But noting that it would be helpful for mothers to begin bonding with their infants after birth is different from claiming that women must bond because sheep do. More instructive than castigating those who learn from animal behavior is realizing that humans, even ancestral humans, would virtually never be in a situation where mistaken identity of an infant is likely to occur; human babies cannot wander off and try to nurse from the neighbor lady down the street. The prolonged contact that does occur between a mother and infant and gradually results in a strong emotional tie suggests instead that people would be reasonably good prospects for adoption, and that there is no magical window for creating a bond between parents or other caregivers and children. It does not dictate adoption; people in different cultures vary in their receptivity to taking on unrelated offspring. But biology can suggest ways in which human behavior is inclined.

Biology also reveals much wider diversity of behavior in animals than we might have first thought possible, and much of that information can be used to break down stereotypes about masculine and feminine behavior. Males can be good caregivers of offspring, and females can be sexually voracious. But we have to be aware of the stereotypes to be able to break them, and this is why I suggested that feminism has more to offer biology than vice versa. Feminism can bring about an awareness of the influence of gender bias on a multitude of activities, and science is not exempted. Discovery of bias, however, should not encourage us to abandon ship.

In a meticulous documentation of contributors to women's lack of progress in attaining social equality, the psychologist Virginia Valian points out that people of both sexes tend to overrate men's achievements while undervaluing those of women. She suggests that these nonconscious generalizations arise from gender schema, hypotheses about how the sexes differ that we develop as we grow up. Such preconceptions have wide ramifications, but they can be illustrated in their most fundamental form by a very simple experiment. Researchers asked college students (one of the model systems for psychology) to estimate the height of people in photographs. The photographs all contained a reference object, such as a desk, which could be used to standardize the estimates. The researchers arranged their models so that the men and women in the photographs had equal height distributions, which should mean that the students' estimates should not differ according to the sex of the model. Instead, estimates of men's height were consistently higher than estimates of women's height, despite the clear available evidence to the contrary.

Valian concludes that gender schema influence our judgments of all manner of things, even those which are supposedly objective. The implication, of course, is that these schema will be at least as powerful and potentially damaging when they are applied to more subjective evaluations, such as the competence of a person in a demanding profession like medicine or the law. My point is slightly different. No one hearing about this example would conclude that we should stop measuring – or even estimating – height in people, nor that feet and inches are not objective measures. Yet those who shun examinations of animal behavior because the behavior is interpreted in sexist ways are doing a very similar thing.

The lesson for us is that feminism, or at least the awareness of gender bias that feminism teaches, has a great deal to tell us about how we can look at other organisms' behavior.

This is not to say, however, that it is straightforward to apply information about animals to humans, as long as one is well enough informed about the animals in question. Even with detailed information about animal lives, we can easily go astray when drawing conclusions about humans. One potential problem is the one I outlined in my discussion of the imprinting and bonding studies, that too little knowledge is a dangerous thing. Two others are common as well, and I briefly detail them below.

SPANDRELS IN THE PLEISTOCENE

As I discussed in the chapter on female orgasm, Richard Lewontin and Stephen Jay Gould have championed a critique of evolutionary explanations of traits, including behavioral traits, by claiming that many so-called adaptations are instead by-products of selection. Their now-classic paper alludes to the spandrels in San Marco's cathedral, structures which arise as a by-product of arch construction but which are used as decorative objects in their own right. It would be misguided to say that spandrels were made so that they could carry gilded motifs. Similarly, the authors suggest that many features of human and animal life that can be rationalized as adaptations are simply neutral occurrences that did not arise through selection.

Defense of adaptationist reasoning has been prolific in the thirty-some years since Gould and Lewontin wrote their paper on the spandrels, and many evolutionary biologists now agree that their attack was a bit overzealous. The criticisms were useful, however, in focusing attention on sloppy reasoning, and in matters of gender such attention is particularly important. It is easy to construct explanations of the adaptive significance of many human and animal sex differences, like the one in spatial ability. The problem is that we know very little about the social environment of the Pleistocene, the favorite scenario-setter for many explanations of human behavior. Did early humans have rigid sex roles, or did men and women have flexible duties, with hunting taking a primary place in some societies and a small role in others? Some attributes are given: females are the sex that gives birth or lays eggs, and males can more often than not increase their reproductive success by increasing the number of females they mate with. But this does not dictate animal behavior, much less that of humans. Furthermore, some of the sex differences we strive to explain turn out to disappear on closer scrutiny or with differences in rearing, like the one in verbal ability. Feminism can keep science honest by ensuring that the phenomena to be explained are not merely sexist spandrels.

Finally, we need to consider alternative explanations for behaviors, again whether in humans or nonhumans. This sounds obvious, but it can be tempting to fall under the tyranny of a "biological" explanation, assuming that such explanations are unitary and unchallenged within science. Any biological explanation must be the right and only explanation, according to this view, so that rejecting any one idea means rejecting the entire principle of biology as relevant to behavior. Such positions are merely polarizing. For example, much is made of the supposed tendency for men to want frequent sex with many different partners, which could be interpreted as an adaptation because of the generalizations I sketched above. A recent study claimed to find support for this idea by asking physically attractive research assistants to proposition strangers on college campuses. Most of the men were willing to go off with unknown women to have sex, while none of the women responded with anything approaching enthusiasm for the idea. The conclusion was that men are, as hypothesized, inherently hornier than women, and by implication male philandering was understandable even when it was not condoned.

But wait a minute. Critics have pointed out an alternative explanation for the findings. Women have been taught since birth that men are potentially dangerous, and that they risk violence when they approach strange men. As Natalie Angier asks, "Could it be that they are in fear of their life rather than uninterested in the pleasure a handsome man might bring them? And could it be that young women just don't scare men physically the way young men do women?" (p. 336). At the very least, it is a plausible alternative, and one that could be tested in its own right. It is a proximate explanation, relying on immediate mechanisms, rather than an ultimate one, relying on adaptive significance. The behavior may in fact represent an adaptation, or it may not. The problem in interpreting the study as it stands is that one cannot do an experiment that controls for the proximate effects of society on human behavior. Alternative hypotheses are crucial in most areas of science, and nowhere more so than in trying to apply principles of evolution to human behavior.

Note that I am not arguing that this criticism negates the likelihood of selection acting on males to encourage their competition for mates. Males in general have less to lose by fertilizing an extra female, even one of relatively poor quality, than females have to lose by having their one set of offspring sired by an inferior mate. Nor am I suggesting that humans are somehow exempt from the evolutionary pressures that influence all other organisms. But these generalizations are colored by the animals we choose as models, by our own biases and preconceptions, and by our skill and resourcefulness in testing the ideas we develop. Looking at bonobos suggests different principles than does looking at sheep. I have suggested in this book that we must be very careful about which animals we use for formulating these principles, and that we cannot choose them because they support or refute pet ideas. The more we study what animals do, the more exceptions, caveats, and constraints emerge to shape the generalities. Feminists cannot abandon all biological explanations of behavior, whether in humans or other animals, because of the worry that some finding detrimental to women will emerge. Doing so would give a power to science that is misplaced. Men and women are not the same, from the standpoint of either physiology or evolution. Neither are male and female goldfish, or fruit flies, or weasels. But this does not mean that men and women do not deserve to be treated equally. At the same time, we can achieve equality without becoming sexual separatists.

Evolution provides the single most rational explanation for every living phenomenon on earth. It helps us understand how organisms are related to one another, how the diversity of life arose, whether a reed warbler female is likely to have any surviving young in a given year, and why diet may help determine the probability of getting heart disease. Suggesting that evolutionary biology is irrelevant to human lives is as foolish as suggesting that it is irrelevant to the lives of fruit flies. Feminists cannot abandon animals to biology and claim no truck with scientists. This damages our credibility and leaves us with a diminished understanding of both other animals and ourselves. Like most relationships, the connection between feminism and biology is filled with the need to simultaneously seek independence and find succor. Neither partner benefits from trying to be subsumed by the other, but severing the ties is counterproductive for both.

SELECTED READINGS

For each chapter I indicate selected readings that contain general information for readers interested in delving further into a topic. In addition, all of the works cited in the text are listed in the References at the end of the book.

INTRODUCTION

Bleier, R., ed. 1986. *Feminist Approaches to Science.* New York: Pergamon Press. A collection of papers generally critical of male bias in science, with some discussion of sociobiology.

Dawkins, R. 1976. *The Selfish Gene.* Oxford: Oxford University Press. The classic examination of how selection acts, not on individuals, but on genes and gene lineages.

Fausto-Sterling, A. 1992. *Myths of Gender: Biological Theories about Women and Men.* 2d ed. New York: Basic Books. A feminist critique of many issues in the biology of sex differences.

Schiebinger, L. 1999. *Has Feminism Changed Science?* Cambridge: Harvard University Press. Considers whether a feminist perspective has changed not only the representation of women in science but the way in which science itself is done.

Segerstråle, U. 2000. *Defenders of the Truth: The Battle for Science in the Sociobiology Debate and Beyond.* Oxford: Oxford University Press. A forthright and careful analysis of the debate between proponents and critics of sociobiology.

Wilson, E. O. 1975. *Sociobiology: The New Synthesis.* Cambridge: Harvard University Press. The classic that started the controversy; most of its chapters are concerned exclusively with animal social behavior, with only the last chapter containing information about humans.

ONE. SEX AND THE DEATH OF A LOON

Clutton-Brock, T. H., F. E. Guinness, and S. D. Albon. 1982. *Red Deer: The Behaviour and Ecology of Two Sexes*. Chicago: University of Chicago Press. A comprehensive study of one of the best-known wild populations of mammals.

Ehrenreich, B., and D. English. 1978. *For Her Own Good: 150 Years of the Experts' Advice to Women*. New York: Anchor Press, Doubleday. A witty discussion of the many ways that women have been told what to do—about their bodies, their children, and their lives—by the prevailing "experts" of the time.

Hrdy, S. B. 1981. *The Woman That Never Evolved*. Cambridge: Harvard University Press. A look at the evolutionary forces acting on females, particularly among primates.

Møller, A. P. 1994. *Sexual Selection and the Barn Swallow*. Oxford: Oxford University Press. Summarizes findings about the reproductive biology of one of the best-studied bird species in the world; written for both specialists and nonscientists.

Tavris, C. 1992. *The Mismeasure of Woman*. New York: Simon & Schuster. A well-written and insightful analysis of how women have been inaccurately held up as better, worse, or abnormal compared with men.

TWO. SUBSTITUTE STEREOTYPES

Adams, C. J., ed. 1993. *Ecofeminism and the Sacred*. New York: Continuum. A general introduction to the ecofeminist movement and its connection to spirituality.

Diamond, I., and G. F. Orenstein, eds. 1990. *Reweaving the World: The Emergence of Ecofeminism*. San Francisco: Sierra Club Books. A collection of essays on the origins and political and social implications of ecofeminism. Contains articles by Ynestra King and Brian Swimme.

Gowaty, P. A. 1997. Principles of females' perspectives in avian behavioral ecology. *Journal of Avian Biology* 28: 95–102. Written by one of the foremost feminist evolutionary biologists, this article is intended for specialists in the field but still gives the flavor of how behavioral scientists critique biology.

Mellor, M. 1997. *Feminism and Ecology*. Cambridge, U.K.: Polity Press. Along with Merchant's book, one of the best-known works on ecofeminism.

Merchant, C. 1996. *Earthcare: Women and the Environment*. New York: Routledge.

Tavris, C. 1992. *The Mismeasure of Woman*. New York: Simon & Schuster. A well-written and insightful analysis of how women have been inaccurately held up as better, worse, or abnormal compared with men.

THREE. SELFLESS MOTHERHOOD

Harlow, H. F., and M. K. Harlow. 1962. Social deprivation in monkeys. *Scientific American* 207 (Nov.):136–146. Describes several of the key experiments demonstrating the role of social experience in the development of monkeys, including some arresting photographs.

Hrdy, S. B. 1999. *Mother Nature.* New York: Pantheon. A thorough and readable account of the evolution of mother-offspring interactions; highly recommended.

FOUR. DNA AND THE MEANING OF MARRIAGE

Alcock, J. 1998. *Animal Behavior.* 7th ed. Sunderland, Mass.: Sinauer Associates. One of the best introductions to many issues in animal behavior, this textbook takes an evolutionary viewpoint.

Bray, O. E., J. J. Kennelly, and J. L. Guarino. 1975. Fertility of eggs produced on territories of vasectomized red-winged blackbirds. *Wilson Bulletin* 87:187–195. Written for a specialist audience, but certainly understandable to the lay reader, this is the paper that hinted at the revolution to come.

Gowaty, P. A. 1997. Principles of females' perspectives in avian behavioral ecology. *Journal of Avian Biology* 28: 95–102. Written by one of the foremost feminist evolutionary biologists, this article is intended for specialists in the field but still gives the flavor of how behavioral scientists critique biology.

Petrie, M., and B. Kempenaers. 1998. Extra-pair paternity in birds: explaining variation between species and populations. *Trends in Ecology and Evolution* 13: 52–58. A good summary of our understanding of extra-pair copulations, written for scientists in areas outside of behavioral biology.

FIVE. THE CARE AND MANAGEMENT OF SPERM

Baker, Robin. 1996. *Sperm Wars: The Science of Sex.* Toronto: HarperCollins. Colorfully written account of sperm competition and its consequences in humans.

Birkhead, T. R., and A. P. Møller, eds. 1998. *Sperm Competition and Sexual Selection.* San Diego and London: Academic Press. In contrast to Baker's book, this is a scholarly collection of articles on sperm competition in a variety of animals. Both the Simmons and Siva-Jothy and the Eberhard chapters are included in it.

Eberhard, W. G. 1996. *Female Control.* Princeton: Princeton University Press. Although written for an academic audience, this book is packed with bizarre and fascinating examples of how females might influence how males fertilize their eggs after copulation.

SIX. SEX AND THE *SCALA NATURAE*

Arnqvist, G. 1998. Comparative evidence for the evolution of genitalia by sexual selection. *Nature* 393: 784–786. Written for a specialist audience, this is nonetheless a good illustration of how information about the behavior or structures of different species can be used to infer evolution.

Eberhard, W. G. 1993. Animal genitalia and female choice. In *Exploring Animal Behavior: Readings from American Scientist,* ed. P. W. Sherman and J. Alcock, 180–187. Sunderland, Mass.: Sinauer Associates, Inc. A more accessible article than Arnqvist's, introducing the idea of cryptic female choice occurring after mating.

Hodos, W., and C. B. G. Campbell. 1969. *Scala Naturae:* why there is no theory in comparative psychology. *Psychological Review* 76:337–350. The classic paper bemoaning the use of the *scala naturae* by scientists.

SEVEN. BONOBOS

Block, S. Web site: www.blockbonobofoundation.org.

De Waal, F. 1997. *Bonobo: The Forgotten Ape.* Berkeley and Los Angeles: University of California Press. A beautifully illustrated introduction to bonobos, by one of the world's experts.

Lilly, J. C. 1961. *Man and Dolphin.* New York: Doubleday. Along with *The Mind of the Dolphin,* a foundation for the modern fascination with dolphins and their mental capabilities.

Lilly, J. C. 1967. *The Mind of the Dolphin.* New York: Doubleday.

Lilly, J. C. Web site: www.garage.co.jp/lilly/thestory.html

Pryor, K., and K. S. Norris. 1991. *Dolphin Societies: Discoveries and Puzzles.* Berkeley and Los Angeles: University of California Press. A popular book by two scientists interested in the natural lives of dolphins.

EIGHT. THE ALPHA CHICKEN

Duffy, M., and K. Tumulty. Gore's secret guru. *Time* magazine, Nov. 8, 1999. Naomi Wolf and the alpha VP.

Ehrenreich, B. 1997. *Blood Rites.* New York: Henry Holt. A social critic's view of how male violence arose and shaped society.

Ghiglieri, M. P. 1999. *The Dark Side of Man.* Reading, Mass.: Perseus Books. Puts forward a rather deterministic view of the origins of human violence.

Huntingford, F., and A. Turner. 1987. *Animal Conflict.* London: Chapman & Hall. Scholarly treatment of how animal conflicts occur in nature and what scientists think they mean.

Lorenz, K. Z. 1952. *King Solomon's Ring.* New York: Thomas Y. Crowell. Still interesting today, Lorenz's book proposes explanations for many aspects of behavior in both humans and animals.

Wrangham, R., and D. Peterson. 1996. *Demonic Males: Apes and the Origins of Human Violence.* Boston: Houghton Mifflin. Anthropologists' views on how human violence may be related to that in other primates.

NINE. SOCCER, ADAPTATION, AND ORGASMS

Alcock, J. 1987. Ardent adaptationism. *Natural History* 4/87: 4 (followed by response from Gould). Along with the Gould response and the note from Sarah Hrdy (below), summarizes the various takes on the adaptive nature of human female orgasm.

Angier, N. 1999. *Woman: An Intimate Geography.* Boston: Houghton Mifflin. A wide-ranging and beautifully written consideration of human females, including but not limited to sexuality.

Gould, S. J. 1987. Freudian slip. *Natural History* 2/87:14–21. More on adaptation and female orgasm.

Hrdy, S. B. 1996. The evolution of female orgasms: logic please but no atavism. *Animal Behaviour* 52:851–852.

Potts, M., and R. Short. 1999. *Ever Since Adam and Eve.* Cambridge: Cambridge University Press. A popular account of sex differences in humans.

Symons, D. 1979. *The Evolution of Human Sexuality.* New York: Oxford University Press. One of the first looks at human sexuality from an evolutionary perspective.

Tavris, C. 1992. *The Mismeasure of Woman.* New York: Simon and Schuster. A well-written and insightful analysis of how women have been inaccurately held up as better, worse, or abnormal compared with men.

TEN. SACRED OR CELLULAR

Grahn, J. 1993. *Blood, Bread, and Roses: How Menstruation Created the World.* Boston: Beacon Press. Along with Owen's book, a "cycle celebrant" work.

Houppert, K. 1999. *The Curse.* New York: Farrar, Straus and Giroux. A critical, sometimes quirky, popular look at menstrual lore.

Martin, E. 1987. *The Woman in the Body.* Boston: Beacon Press. Feminist critique of issues surrounding women's bodies.

Owen, L. 1993. *Her Blood Is Gold: Celebrating the Power of Menstruation.* San Francisco: Harper.

Profet, M. 1993. Menstruation as a defense against pathogens transported by sperm. *Quarterly Review of Biology* 68:335–386. The original proposal for an adaptive function of menstruation; intended for biologists but understandable to a general reader as well.

Tavris, C. 1992. *The Mismeasure of Woman.* New York: Simon and Schuster. A well-written and insightful analysis of how women have been inaccurately held up as better, worse, or abnormal compared with men.

ELEVEN. THAT'S NOT SEX

Bagemihl, B. 1999. *Biological Exuberance: Animal Homosexuality and Natural Diversity.* New York: St. Martin's Press. A thorough treatise of purported instances of homosexuality among animals; proposes a somewhat New Age explanation for its occurrence.

Burr, C. 1996. *A Separate Creation.* New York: Hyperion. Excellent discussion of the various findings about the biological basis of sexual orientation, written by a journalist.

Hamer, D., and P. Copeland. 1994. *The Science of Desire.* New York: Simon and Schuster. Hamer's own view of his work on the basis of homosexuality.

LeVay, S. 1996. *Queer Science: The Use and Abuse of Research into Homosexuality.* Cambridge: MIT Press. LeVay's treatment of research in sexual orientation, emphasizing his own lab's work on the brain.

TWELVE. CAN VOLES DO MATH?

Alcock, J. 1998. *Animal Behavior.* 7th ed. Sunderland, Mass.: Sinauer. One of the best introductions to many issues in animal behavior, this textbook takes an evolutionary viewpoint.

Anon. 1986. Mathematical genius: in the hormones? *New Scientist,* 29 May. A popular summary and speculation about sex differences in mathematical ability.

Benbow, C. P., and J. C. Stanley. 1980. Sex differences in mathematical ability: fact or artifact? *Science* 210:1262–1264. The original article that sparked much of the controversy about boys and girls and math.

Maraffi, M. Girls' attitudes, self-expectations, and performance in math: an annotated bibliography. From Math Forum web site: http://forum.swarthmore.edu/sarah/Discussion.Sessions/biblio.attitudes.html. Contains many useful articles on mathematics education and girls.

CONCLUSION

Angier, N. 1999. *Woman: An Intimate Geography.* Boston: Houghton Mifflin. A wide-ranging and beautifully written consideration of human females, including but not limited to sexuality.

Hrdy, S. B. 1999. *Mother Nature.* New York: Pantheon. A thorough and readable account of the evolution of mother-offspring interactions; highly recommended.

Valian, V. 1998. *Why So Slow?* Cambridge: MIT Press. Thoughtful discussion of why women's advancement in many professions still lags behind that of men.

REFERENCES

Abramson, P. R., and S. D. Pinkerton, eds. 1995. *Sexual Nature, Sexual Culture.* Chicago: University of Chicago Press.

Alcock, J. 1987. Ardent adaptationism. *Natural History* 4/87: 4.

Angier, N. 1999. *Woman: An Intimate Geography.* Boston: Houghton Mifflin.

Archer, J. 1988. *The Behavioural Biology of Aggression.* Cambridge: Cambridge University Press.

Arnqvist, G. 1998. Comparative evidence for the evolution of genitalia by sexual selection. *Nature* 393: 784–786.

Bagemihl, B. 1999. *Biological Exuberance: Animal Homosexuality and Natural Diversity.* New York: St. Martin's Press.

Bailey, J. M., and R. C. Pillard. 1991. A genetic study of male sexual orientation. *Archives of General Psychiatry* 48:1089–96.

———. 1993. A genetic study of male sexual orientation (letter). *Archives of General Psychiatry* 50:240–241.

Bajema, C. J., ed. 1984. *Evolution by Sexual Selection Theory prior to 1900.* New York: Van Nostrand Reinhold.

Baker, R. R., and M. A. Bellis. 1995. *Human Sperm Competition.* London: Chapman & Hall.

Benbow, C. P., and J. C. Stanley. 1980. Sex differences in mathematical ability: fact or artifact? *Science* 210:1262–1264.

———. 1983. Sex differences in mathematical reasoning ability: more facts. *Science* 222:1029–1031.

Berenbaum, M. R. 1995. *Bugs in the System: Insects and Their Impact on Human Affairs.* Reading, Mass.: Perseus Books.

Beston, H. 1928. *The Outermost House.* New York: Doubleday.

Birkhead, T. R., and A. P. Møller, eds. 1992. *Sperm Competition in Birds: Evolutionary Causes and Consequences.* New York: Academic Press.

Bray, O. E., J. J. Kennelly, and J. L Guarino. 1975. Fertility of eggs produced on territories of vasectomized red-winged blackbirds. *Wilson Bulletin* 87:187–195.

Bright, S. 1998. *Susie Sexpert's Lesbian Sex World.* 2d ed. San Francisco: Cleis Press.

Brooke, M. D., N. B. Davies, and D. G. Noble. 1998. Rapid decline of host defences in response to reduced cuckoo parasitism: behavioural flexibility of reed warblers in a changing world. *Proceedings of the Royal Society of London,* Ser. B, 265:1277–1282.

Broverman, I. K., D. M. Broverman, and F. E. Clarkson. 1970. Sex-role stereotypes and clinical judgements of mental health. *Journal of Consulting Clinical Psychology* 34:1–7.

Buckley, T., and A. Gottlieb, eds. 1988. *Blood Magic: The Anthropology of Menstruation.* Berkeley and Los Angeles: University of California Press.

Burley, N. T., P. G. Parker, and K. Lundy. 1996. Sexual selection and extrapair fertilization in a socially monogamous passerine, the zebra finch *(Taeniopygia guttata). Behavioral Ecology* 7:218–226.

Clarke, J. 1994. The meaning of menstruation in the elimination of abnormal embryos. *Human Reproduction* 9:1204–1207.

Clutton-Brock, T. H., S. D. Albon, and F. E. Guinness. 1986. Great expectations—dominance, breeding success and offspring sex-ratios in red deer. *Animal Behaviour* 34:460–471.

Clutton-Brock, T. H., F. E. Guinness, and S. D. Albon. 1982. *Red Deer: The Behaviour and Ecology of Two Sexes.* Chicago: University of Chicago Press.

Connor, R. C., and D. M. Peterson. 1994. *The Lives of Whales and Dolphins.* New York: Henry Holt.

Coveney, L., M. Jackson, S. Jeffreys, L. Kaye, and P. Mahoney. 1984. *The Sexuality Papers.* London: Hutchison.

Darwin, C. 1871. *The Descent of Man and Selection in Relation to Sex.* New York: Modern Library reissue.

Davies, N. B. 1999. Cuckoos and cowbirds versus hosts: co-evolutionary lag and equilibrium. *Ostrich* 70:71–79.

Dawkins, R. 1979. Twelve misunderstandings of kin selection. *Zeitschrift für Tierpsychologie* 51:184–200.

Devine, E., and M. Clark. 1967. *The Dolphin Smile.* New York: Macmillan.

De Waal, F. 1997. *Bonobo: The Forgotten Ape.* Berkeley and Los Angeles: University of California Press.

Doak, W. 1989. *Encounters with Whales and Dolphins.* New York: Sheridan.

Dreifus, C. 1999. Going ape. *Ms.* 9(5):48–54.

Eals, M., and I. Silverman. 1994. The hunter-gatherer theory of spatial sex differences: proximate factors mediating the female advantage in recall of object arrays. *Ethology and Sociobiology* 15:95–105.

Eberhard, W. G. 1996. *Female Control*. Princeton: Princeton University Press.

———. 1998. Female roles in sperm competition. In *Sperm Competition and Sexual Selection,* ed. T. R. Birkhead and A. P. Møller. San Diego and London: Academic Press.

Eberhard, W. G., B. A. Huber, R. L. Rodriguez, R. D. Briceno, I. Salas, and V. Rodriquez. 1998. One size fits all? Relationships between the size and degree of variation in genitalia and other body parts in twenty species of insects and spiders. *Evolution* 52:415–431.

Ehrenreich, B. 1997. *Blood Rites*. New York: Henry Holt.

Ehrenreich, B., and D. English. 1978. *For Her Own Good: 150 Years of the Experts' Advice to Women*. New York: Anchor Press, Doubleday.

Ehrlich, P. R. and A. Ehrlich. 1981. *Extinction: The Causes and Consequences of the Disappearance of Species*. New York: Random House.

Ellis, L., and L. Ebertz, eds. 1997. *Sexual Orientation: Toward Biological Understanding*. Westport, Conn.: Praeger.

Emlen, S. T. 1995. Can avian biology be useful to the social sciences? *Journal of Avian Biology* 26:273–276.

———. 1995. An evolutionary theory of the family. *Proceedings of the National Academy of Sciences USA* 92:8092–8099.

Ethington, C. A., and L. M. Wolfle. 1986. A structural model of mathematics achievement for men and women. *American Educational Research Journal* 23: 65–75.

Evans, P. G. H. 1987. *The Natural History of Whales and Dolphins*. London: Christopher Helm.

Faludi, S. 1991. *Backlash: The Undeclared War against American Women*. New York: Crown.

Fedigan, L. M. 1982. *Primate paradigms: sex roles and social bonds*. Montreal: Eden Press.

———. 1986. The changing role of women in models of human evolution. *Annual Review of Anthropology* 15: 25–66.

Fennema, E., and L. E. Hart. 1994. Gender and the JRME. *Journal of Research in Mathematics Education* 25:648–659.

Fennema, E., and J. A. Sherman. 1977. Sexual stereotyping and mathematics learning. *Arithmetic Teacher* 24:369–372.

Finn, C. A. 1987. Why do women and some other primates menstruate? *Perspectives in Biology and Medicine* 30:566–574.

———. 1994. The meaning of menstruation. *Human Reproduction* 9:1202–1204.

———. 1996. Why do women menstruate? historical and evolutionary review. *European Journal of Obstetrics & Gynecology and Reproductive Biology* 70:3–8.

French, J. A. 1997. Proximate regulation of singular breeding in callitrichid primates. In *Cooperative Breeding in Mammals,* ed. N. G. Solomon and J. A. French, 34–75. Cambridge: Cambridge University Press.

Galea, L. A. M., M. Kavaliers, and K.-P. Ossenkopp. 1996. Sexually dimorphic spatial learning in meadow voles *Microtus pennsylvanicus* and deer mice *Peromyscus maniculatus. Journal of Experimental Biology* 199:195–200.

Gaulin, S. J. C., and R. W. FitzGerald. 1986. Sex differences in spatial ability: an evolutionary hypothesis and test. *American Naturalist* 127:74–88.

Ghiglieri, M. P. 1999. *The Dark Side of Man.* Reading, Mass.: Perseus Books.

Gibbs, H. L., P. J. Weatherhead, P. T. Boag, B. N. White, L. M. Tabak, and D. J. Hoysak. 1990. Realized reproductive success of polygynous red-winged blackbirds revealed by DNA markers. *Science* 250: (4986) 1394–1397.

Gould, S. J., and R. C. Lewontin. 1979. The spandrels of San Marco and the Panglossian paradigm: a critique of the adaptationist programme. *Proceedings of the Royal Society, London,* Ser. B, 205:581–598.

Gowaty, P. A. 1994. Architects of sperm competition. *Trends in Ecology and Evolution* 9:160–162.

———. 1997. Principles of females' perspectives in avian behavioral ecology. *Journal of Avian Biology* 28: 95–102.

Gowaty, P. A., and N. Buschhaus. 1998. Ultimate causation of aggressive and forced copulation in birds: female resistance, the CODE hypothesis, and social monogamy. *American Zoologist* 38:207–225.

Grant, E., ed. 1974. *A Source Book in Medieval Science.* Cambridge, Mass.: Harvard University Press.

Gray, J. A., and A. W. H. Buffery. 1971. Sex differences in emotional and cognitive behaviour in mammals including man: adaptive and neural bases. *Acta Psychologica* 35:89–111.

Gray, R. 1997. "In the belly of the monster": feminism, developmental systems, and evolutionary explanations. In *Feminism and Evolutionary Biology,* ed. P. A. Gowaty, 385–413. New York: Chapman and Hall.

Gyllenstein, U. B., S. Jakobsson, and H. Temrin. 1990. No evidence for illegitimate young in monogamous and polygynous warblers. *Nature* 343:168–170.

Hager, L. D., ed. 1997. *Women in Human Evolution.* London: Routledge.

Haig, D. 1999. Genetic conflicts of pregnancy and childhood. In *Evolution in Health and Disease,* ed. S. C. Stearns. Oxford: Oxford University Press.

Hamer, D., and P. Copeland. 1994. *The Science of Desire.* New York: Simon and Schuster.

Hamilton, W. D. 1979. Wingless and fighting males in fig wasps and other insects. In *Reproductive Competition, Mate Choice and Sexual Selection in Insects,* ed. M. S. Blum and N. A. Blum. New York: Academic Press.

Hansson, B., S. Bensch, and D. Hasselquist. 1997. Infanticide in great reed warblers: secondary females destroy eggs of primary females. *Animal Behaviour* 54: 297–304.

Harlow, H. F., and M. K. Harlow. 1962. Social deprivation in monkeys. *Scientific American* 207 (Nov.):136–146.

Harlow, H. F., and R. R. Zimmerman. 1959. Affectional responses in the infant monkey. *Science* 130:421–432.

Hart, S. 1996. *The Language of Animals*. New York: Henry Holt.

Healy, S. D., S. R. Braham, and V. A. Braithwaite. 1999. Spatial working memory in rats: no difference between the sexes. *Proceedings of the Royal Society, London,* Ser. B, 266:2303–2308.

Henneberger, M. 1999. Naomi Wolf, feminist consultant to Gore, clarifies her campaign role. *New York Times,* Nov. 5.

Herdt, G. 1997. *Same Sex, Different Cultures.* Boulder, Colo.: Westview Press.

Hodos, W., and C. B. G. Campbell. 1969. *Scala Naturae:* why there is no theory in comparative psychology. *Psychological Review* 76:337–350.

Houppert, K. 1999. *The Curse.* New York: Farrar, Straus and Giroux.

Housman, A. E. 1988. *Last poems.* Reprint edition, London: Amereon.

Hrdy, S. B. 1981. *The Woman That Never Evolved.* Cambridge: Harvard University Press.

———. 1986. Empathy, polyandry, and the myth of the coy female. In *Feminist Approaches to Science,* ed. R. Bleier. New York: Pergamon Press.

———. 1996. The evolution of female orgasms: logic please but no atavism. *Animal Behaviour* 52:851–852.

———. 1999. *Mother Nature.* New York: Pantheon.

Jacobs, L. F. 1996. Sexual selection and the brain. *Trends in Ecology and Evolution* 11:82–86.

Jennions, M. D., and M. Petrie. 2000. Why do females mate multiply? A review of the genetic benefits. *Biological Reviews* 75:21–64.

Kaitala, A. 1996. Oviposition on the back of conspecifics: an unusual reproductive tactic in a coreid bug. *Oikos* 77:381–389.

———. 1998. Is egg carrying attractive? Mate choice in the golden egg bug (Coreidae, Heteroptera). *Proceedings of the Royal Society of London,* Ser. B, 65:779–783.

———. 1999. Counterstrategy to egg dumping in a Coreid bug: recipient individuals discard eggs from their backs. *Journal of Insect Behavior* 12:225–232.

Kaitala, A., and A. H. Axen. 2000. Egg load and mating status of the golden egg bug affect predation risk. *Ecology* 81:876–880.

Kano, T. 1992. *The Last Ape: Pygmy Chimpanzee Behavior and Ecology.* Stanford: Stanford University Press.

Keller, E. F. 1985. *Reflections on Gender and Science.* New Haven: Yale University Press.

King, Y. 1990. Healing the wounds: feminism, ecology, and the nature/culture dualism. In *Reweaving the World: The Emergence of Ecofeminism,* ed. I. Diamond and G. F. Orenstein. San Francisco: Sierra Club Books.

Knight, C. 1991. *Blood Relations: Menstruation and the Origins of Culture.* New Haven: Yale University Press.

Koenig, W. D., J. Haydock, and M. T. Stanback. 1998. Reproductive roles in the cooperatively breeding acorn woodpecker: incest avoidance versus reproductive competition. *American Naturalist* 151:243–255.

Koenig, W. D., and R. L. Mumme. 1987. *Population Ecology of the Cooperatively Breeding Acorn Woodpecker.* Princeton: Princeton University Press.

Konner, M. 1982. *The Tangled Wing: Biological Constraints on the Human Spirit.* New York: Holt, Rinehart, and Winston.

———. 1988. Is orgasm essential? *The Sciences* 28(2):4–7.

Kruuk, L. E. B., T. H. Clutton-Brock, S. D. Albon, J. M. Pemberton, and F. E. Guinness. 1999. Population density affects sex ratio variation in red deer. *Nature* 399:459–461.

Kummer, H. 1971. *Primate Societies: Group Techniques of Ecological Adaptation.* Chicago: Aldine.

Lawton, M. F., W. R. Garstka, and J. C. Hanks. 1997. The mask of theory and the face of nature. In *Feminism and Evolutionary Biology,* ed. P. A. Gowaty. New York: Chapman and Hall.

LeVay, S. 1991. A difference in hypothalamic structure between heterosexual and homosexual men. *Science* 253:1034–1037.

———. 1996. *Queer Science: The Use and Abuse of Research into Homosexuality.* Cambridge: MIT Press.

Lilly, J. C. 1961. *Man and Dolphin.* New York: Doubleday.

———. 1967. *The Mind of the Dolphin.* New York: Doubleday.

Lorenz, K. Z. 1952. *King Solomon's Ring.* New York: Thomas Y. Crowell.

Margulis, L., and D. Sagan. 1991. *Mystery Dance: On the Evolution of Human Sexuality.* New York: Summit Books.

Martin, E. 1987. *The Woman in the Body.* Boston: Beacon Press.

Marzluff, J. M., and R. P. Balda. 1992. *The Pinyon Jay: Behavioral Ecology of a Colonial and Cooperative Corvid.* London: Poyser.

Mason, W. A., and S. P. Mendoza, eds. 1993. *Primate Social Conflict.* Albany: SUNY Press.

Maynard Smith, J., and M. G. Ridpath. 1972. Wife sharing in the Tasmanian native hen, *Tribonyx mortierii:* a case of kin selection? *American Naturalist* 106:447–452.

McCarthy, S. 1999. The fabulous kingdom of gay animals. Ivory Tower, *Salon Magazine* 3 (http://www.salon.com/it/feature/1999/03/cov_15featurea.html).

McClintock, M. K. 1971. Menstrual synchrony and suppression. *Nature* 229:244–245.

McIntyre, J., ed. 1974. *Mind in the Waters.* New York: Charles Scribner's Sons.

McKnight, J. 1997. *Straight Science? Homosexuality, Evolution and Adaptation.* New York: Routledge.

Mellor, M. 1997. *Feminism and Ecology.* Cambridge, U.K.: Polity Press.

Merchant, C. 1996. *Earthcare: Women and the Environment.* New York: Routledge.

Meyer, J. S., M. A. Novak, R. E. Bowman, and H. F. Harlow. 1975. Behavioral and hormonal effects of attachment object separation in surrogate-peer-reared and mother-reared infant rhesus monkeys. *Developmental Psychobiology* 8: 425–435.

Møller, A. P. 1994. *Sexual Selection and the Barn Swallow.* Oxford: Oxford University Press.

Morgen, S., ed. 1989. *Gender and Anthropology: Critical Reviews for Research and Teaching.* Washington, D.C.: American Anthropological Association.

Morse, M. 1995. *Women Changing Science: Voices from a Field in Transition.* New York: Insight Books.

Murchison, C. A., ed. 1935. *A Handbook of Social Psychology.* New York: Clark University Press.

Ness, E. 1999. Sin County Almanac. *Grist* online magazine (www.gristmagazine.com/grist/limb/limb093099).

Orians, G. H. 1969. On the evolution of mating systems in birds and mammals. *American Naturalist* 103:589–603.

Owen, L. 1993. *Her Blood Is Gold: Celebrating the Power of Menstruation.* San Francisco: Harper.

Parish, A. R. Female relationships in bonobos *(Pan paniscus):* evidence for bonding, cooperation, and female dominance in a male-philopatric species. *Human Nature* 7:61–96.

Parker, G. A. 1970. Sperm competition and its evolutionary consequences in the insects. *Biology Reviews* 45:525–567.

Payne, R. 1995. *Among Whales.* New York: Scribner.

Petrie, M., C. Doums, and A. P. Møller. 1998. The degree of extra-pair paternity increases with genetic variability. *Proceedings of the National Academy of Sciences USA* 95:9390–9395.

Petrie, M., and B. Kempenaers. 1998. Extra-pair paternity in birds: explaining variation between species and populations. *Trends in Ecology and Evolution* 13: 52–58.

Potts, M., and R. Short. 1999. *Ever Since Adam and Eve.* Cambridge: Cambridge University Press.

Profet, M. 1993. Menstruation as a defense against pathogens transported by sperm. *Quarterly Review of Biology* 68:335–386.

Rosario, V. A., ed. 1997. *Science and Homosexualities.* New York: Routledge.

Rosen, R. C., and J. G. Beck. 1988. *Patterns of Sexual Arousal: Psychophysiological Processes and Clinical Applications.* New York: Guilford Press.

Ruppenthal, G. C., G. L. Arling, H. F. Harlow, G. P. Sackett, and S. J. Suomi. 1976. 10-year perspective of motherless-mother monkey behavior. *Journal of Abnormal Psychology* 85:341–349.

Sapolsky, R. M., S. C. Alberts, and J. Altmann. 1997. Hypercortisolism associated with social subordinance or social isolation among wild baboons. *Archives of General Psychiatry* 54: 1137–1143.

Sapolsky, R. M., and E. M. Spencer. 1997. Insulin-like growth factor I is suppressed in socially subordinate male baboons. *American Journal of Physiology* 273: R1346–1351.

Savage-Rumbaugh, S., S. G. Shanker, and T. J. Taylor. 1998. *Apes, Language, and the Human Mind.* New York: Oxford University Press.

Schein, M. W., ed. 1975. *Social Hierarchy and Dominance.* Stroudsburg, Pa: Halsted Press.

Schjelderup-Ebbe, T. 1922. Contributions to the social psychology of the domestic chicken. *Zeitschrift fur Tierpsychologie* 88:225–252. Reprinted in *Social Hierarchy and Dominance*, ed. M. W. Schein. 1975. Stroudsburg, Pa.: Halsted Press.

——. 1935. Social behavior in birds. In *A Handbook of Social Psychology,* ed. C. Murchison. New York: Russell and Russell.

Schnierla, T. C. 1946. Problems in the biopsychology of social organization. *Journal of Abnormal Social Psychology* 41:395–397.

Schubert, G., and R. D. Masters, eds. 1991. *Primate Politics.* Carbondale: Southern Illinois University Press.

Secada, W. G., E. Fennema, and L. B. Adajian, eds. 1995. *New Directions for Equity in Mathematics Education.* Cambridge: Cambridge University Press.

Seelye, K. Q. 1999. Adviser pushes Gore to be leader of the pack. *New York Times,* Nov. 1.

Segerstråle, U. 2000. *Defenders of the Truth: The Battle for Science in the Sociobiology Debate and Beyond.* Oxford: Oxford University Press.

Sheldon, B. C. 1994. Sperm competition in the chaffinch: the role of the female. *Animal Behaviour* 47:163–173.

Sheldon, B. C., and T. R. Birkhead. 1994. Reproductive anatomy of the chaffinch in relation to sperm competition. *Condor* 96:1099–1103.

Sherry, D. F., and E. Hampson. 1997. Evolution and the hormonal control of sexually-dimorphic spatial abilities in humans. *Trends in the Cognitive Sciences* 1: 50–56.

Simmons, L. W., and M. T. Siva-Jothy. 1998. Sperm competition in insects: mechanisms and the potential for selection. In *Sperm Competition and Sexual Selection,* ed. T. R. Birkhead and A. P. Møller. San Diego and London: Academic Press.

Slob, A. K., W. H. Groeneveld, and J. J. van der Werf ten Bosch. 1986. Physiological changes during copulation in male and female stumptail macaques *(Macaca arctoides). Physiology and Behavior* 38:891–895.

Smith, R. L., ed. 1984. *Sperm Competition and the Evolution of Animal Mating Systems.* Orlando, Fla.: Academic Press.

Stanford, C. B. 1998. The social behavior of chimpanzees and bonobos. *Current Anthropology* 39:399–420.

Stearns, S. C., ed. 1999. *Evolution in Health and Disease.* Oxford: Oxford University Press.

Strassmann, B. I. 1996. Menstrual hut visits by Dogon women: a hormonal test distinguishes deceit from honest signaling. *Behavioral Ecology* 7:304–315.

——. 1996. The evolution of endometrial cycles and menstruation. *Quarterly Review of Biology* 71:181–220.

——. 1997. The biology of menstruation in *Homo sapiens:* total lifetime menses, fecundity, and nonsynchrony in a natural-fertility population. *Current Anthropology* 38:123–129.

Swimme, B. 1990. How to heal a lobotomy. In *Reweaving the World: The Emergence of Ecofeminism,* ed. I. Diamond, and G. F. Orenstein. San Francisco: Sierra Club Books.

Symons, D. 1979. *The Evolution of Human Sexuality*. New York: Oxford University Press.

Tavris, C. 1992. *The Mismeasure of Woman*. New York: Simon and Schuster.

Thorndike, E. L. 1898. Animal intelligence; an experimental study of the associative processes in animals. *Psychological Review Monograph Supplement* 2:1–109.

Thornhill, R., and S. W. Gangestad. 1996. Human female copulatory orgasm: a human adaptation or phylogenetic holdover. *Animal Behaviour* 52:853–855.

Troisi, A., and M. Carosi. 1998. Female orgasm rate increases with male dominance in Japanese macaques. *Animal Behaviour* 56:1261–1266.

Valian, V. 1998. *Why So Slow?* Cambridge: MIT Press.

Voyer, D., S. Voyer, and M. P. Bryden. 1995. Magnitude of sex differences in spatial abilities: a meta-analysis and consideration of critical variables. *Psychology Bulletin* 117:250–270.

Waage, J. K. 1979. Dual function of the damselfly penis: sperm removal and transfer. *Science* 203:916–918.

Wade, N. 1999. What's it all about, alpha? *New York Times*, Nov. 7.

Wallace, A. R. 1889. *Darwinism: An Exposition of the Theory of Natural Selection with Some of Its Applications*. London: Macmillan.

Wallen, K. 1995. The evolution of female desire. In *Sexual Nature, Sexual Culture*, ed. P. R. Abramson and S. D. Pinkerton. Chicago: University of Chicago Press.

West-Eberhard, M.J. 1979. Sexual selection, social competition, and evolution. Proceedings of the American Philosophical Society 124:222–234.

Wilson, E. O. 1975. *Sociobiology: The New Synthesis*. Cambridge: Harvard University Press.

———. 1984. *Biophilia*. Cambridge: Harvard University Press.

Wrangham, R., and D. Peterson. 1996. *Demonic Males: Apes and the Origins of Human Violence*. Boston: Houghton Mifflin.

INDEX

Rats, 25, 196

Red deer *(Cervus elaphus)*, 25–26

Red jungle fowl *(Gallus gallus)*, 95, 125

Red-necked phalarope *(Phalaropus lobatus)*, 100–101

Red-winged blackbird *(Agelaius phoeniceus)*: DNA testing of chicks, 68–69; mating system of, 63–65; mating territories of, 61–62

Reed warbler *(Acrocephalus scrirpaceus)*, 59–60

Reproductive skew theory, 59

Reproductive success: of altruistic honeybees, 57–58; of brood parasitism, 59–60; of cooperative breeding, 58–59; dominance linked to, 128–30, 131–33; female choice component of, 6–7, 8–9; and homosexuality, 179–80; of infanticidal males, 142–43; male orgasm and, 140, 147–48, 151–52; menstruation as tool of, 164–65; parental care linked to, 55–57; polygyny threshold of, 63; and sex ratio of offspring, 26–27; sexual pleasure versus, 181–83. *See also* Extra-pair copulations; Sperm competition

Reproductive suppression, 128–30

Reweaving the World (Swimme), 46

Rhesus macaques *(Macaca mulatta)*, 48–49, 99

Rhinoceros beetles, 6

Rhum island (Scotland), 26

Richner, Heinz, 54

Roughtoothed dolphin *(Steno bredanensis)*, 109–10

Sagan, Dorion, 146

Salon.com, 177

Sandpipers, 100–102

Sapolsky, Robert, 126

Satyridae butterflies, 80

Savage-Rumbaugh, Sue, 117–18

Scala naturae (scale of nature): abandonment of, 103, 106; balance of nature focus of, 112; defined, 16; vs. phylogenetic tree, 96–97; ranking system of, 94–96, 97–98, 113; sexual behavior research using, 99–100

"*Scala naturae:* Why There Is No Theory in Comparative Psychology" (Hodos and Campbell), 94–95

Schjelderup-Ebbe, Thorleif, 23, 122–23, 127

Schnierla, T. C., 125

Science (journal), 174, 187

The Science of Desire (Hamer and Copeland), 171

Seahorses, 53, 99

Secondary sexual characters: defined, 6; evolution of, and female choice, 6–7, 8–9, 11, 85–86

Segerstråle, Ullica, 4, 11, 14

Semi-altricial species, 206

Sexual behavior: Arnqvist's competing theories on, 104–5; as biologically intuitive, 22; of bonobos, 108, 114, 115–17; of cetaceans, 113–14; dominance relations in, 127–30, 133–34; evolutionary role of, 5; genetics research on, 171–74; hunting linked to, 133–34; phylogeny of, in shorebirds, 100–103; for pleasure vs. reproduction, 181–83; rats as model of, 25; of rhesus macaques, 49; *scala naturae* examination of, 99–100; of sexually monomorphic species, 176. *See also* Extrapair copulations; Homosexuality; Reproductive success; Sperm competition

Sexual imprinting, 207–8

Sexually transmitted diseases in birds, 72–73

Sexual selection: common misconceptions about, 10–14; cryptic female choice theory of, 86–89; Darwinian theory of, 5–7; in evolution of genitalia, 104–5; without *scala naturae*, 106; scientific disinterest in, 7–8; Trivers's formulation of, 8–9

Sheldon, Ben, 79

Sherry, David, 193–94

Shorebirds, 100–103

Short, Roger, 149–50, 151

Shrews, 25, 158

Sialia sialis (eastern bluebird), 30–31, 38, 41

Simmons, Leigh, 83

Simpson, George Gaylord, 7–8

"Sin County Almanac" (Ness), 116

Indexer:	Pat Deminna
Compositor:	Binghamton Valley Composition, LLC
Text:	11.25/13.5 Adobe Garamond
Display:	Perpetua and Adobe Garamond
Printer and binder:	Haddon Craftsmen